THE NEW GEOCENTRISTS

THE NEW GEOCENTRISTS

KARL KEATING

RASSELAS
HOUSE

Copyright © 2015 Karl Keating

Published by Rasselas House
San Diego, California
RasselasHouse.com

Cover design by Damonza.com

ISBN 978-1-942596-00-4

Nonsense can be defended but by nonsense.

—Samuel Johnson, *Life of Johnson*

False ideas may be refuted indeed by argument, but by true ideas alone are they expelled.

—John Henry Newman, *Apologia Pro Vita Sua*

You say that we go round the Sun. If we went round the Moon it would not make a pennyworth of difference to me or to my work.

—Sherlock Holmes, *A Study in Scarlett*

CONTENTS

PART 4

PART 5

PART 6

PREFACE

WHEN I WAS a boy, I collected and read nearly all of Edgar Rice Burroughs' books. Best known as the creator of Tarzan, he wrote more than seventy works of adventure and fantasy. Most of them appeared in series. The books printed during his lifetime included twenty-two in the Tarzan series, ten in the Mars series, four in the Venus series, and six in the Pellucidar series. It was this last series I found most intriguing in its setting. It is one thing to set stories on Mars or Venus; it is something else to set them within the Earth.

The conceit of the Pellucidar books is that the Earth is hollow, its crust being only five hundred miles thick, making for a concave world about seven thousand miles across. In the center of the internal sky is a tiny star that gives perpetual light to the land of Pellucidar, which knows no night. Sight is not lost at a distant horizon because there is no horizon. The land curves up, not down, and sight is lost only in far-off mists or when the up-sloping terrain gets lost in the brilliance of the star.

In Burroughs' telling, Earth's interior land surface is

greater than its exterior land surface: where there is land on the outer surface, there is water on the inner, and where there is water on the outer surface, there is land on the inner. This results in large continents with plenty of room for the wide variety of creatures with which Burroughs liked to inhabit his worlds.

The first book in the Pellucidar series, *At the Earth's Core*, was published in 1914. In that year there still were largely unexplored regions of the world, chiefly in the Arctic and Antarctic. This allowed Burroughs to provide for entrances to Pellucidar from the outer surface—large holes at the poles—and to give the reader an excuse for his suspension of disbelief. Burroughs' schematic would not be so easily accepted by today's reader, given that he is likely to know that every square inch of the globe has been mapped and that a large entrance to Pellucidar, if such an entrance existed, would have been discovered decades ago. In 1914, a writer still could get away with such a setup, even if barely.

Fast forward half a century. Imagine my surprise to discover someone who actually believes that the Earth is hollow. This is the message presented in a booklet called *A Hollow Earth? The Bible Says Yes!* The author is Jim Wilhelmsen, who appears to be a Fundamentalist Protestant. His opening lines give the flavor of the rest of his booklet:

> Modern science will tell you that it is impossible for the Earth to be "hollow." This same modern science will tell you man came from a monkey and there are no such things as UFOs. The "official" position of the Church (Holy Roman version) was that the

Earth was flat. The emerging sciences began to realize that this was wrong. Today in the minds of too many people, the Bible has lost its validity because men were wrong! It is funny that the Bible declared that the Earth was round long before there was a modern science or the Church (Holy Roman type).

There is no reason to discuss Wilhelmsen's arguments for a hollow Earth. They are risible. He seems sincere, yet he is sincerely wrong. He is convinced that his interpretation of Scripture is correct and that the sacred text, if read with open eyes, will reveal the truth of an interior world. It will not. He has misunderstood and misapplied Scripture, though he is not conscious of doing so. He has seen in the Bible geographical and cosmological proofs that simply aren't there.

In these attitudes he is like others who find themselves compelled, chiefly through their interpretations of Scripture, to promote a scientific theory thought ludicrous by nearly all of their contemporaries. These others, unlike Wilhelmsen, have some history to work from and arguments that, though weak, are not always obviously so. Even their biblical exegesis is better laid than his. Yet, for all their relative sophistication, they too are wrong.

Their wrongness is more consequential than his because they have a wider following than he ever could hope to have, and their theory is incorporated into a theological and cultural critique that has no analogue in his system. He has little chance of bringing anyone to his side. They have such a chance, and they want to take advantage of it.

They are the new geocentrists.

WHAT THIS BOOK IS AND ISN'T

THIS BOOK IS titled *The New Geocentrists*, not *The New Geocentrism*. The focus is more on the people than on their scientific and religious claims. Their claims are not neglected, but I make no attempt at comprehensiveness and feel no obligation to do so. Today's geocentrists have worked up dozens of main and hundreds of subsidiary arguments spread over thousands of book pages and innumerable web pages. It would be as tedious to refute those arguments, one by one, as it is to read them—more tedious, actually, since it often requires a full paragraph to correct a single wrongheaded sentence, and there are many wrongheaded sentences in the works of the new geocentrists.

Geocentrists are oblivious to problems in what they write. They are confident in their science and exegesis—confident to the point of not seeing weaknesses in their position. They think their logical armor has no chinks. They betray no doubts, no signs of hesitancy. Conflicting evidence does not exist or is dealt with summarily. The only logical flaws are with those who fail to accept the geocentric thesis.

That thesis is gaining adherents. Its proponents string

together innumerable names, dates, equations, and assertions. Can advocates who have so much to say and who say it so insistently be wrong? More and more people are answering in the negative. They find the cosmological and scriptural ideas put before them by geocentrists to be convincing, even compelling. Flaws in the geocentrist argument are seldom obvious to the intended audience, which is not the scientist or the theologian but the layman, particularly the layman who is susceptible to sweeping claims and to hints of conspiracies.

Those who are coming to accept the geocentric thesis are not doing so because they can follow the physics and can work through the mathematics. They take the representations of the new geocentrists on faith. They think them to be reliable exponents of the facts and trustworthy interpreters of the meanings behind the facts, and they give them extra points for speaking in opposition to settled opinion.

This last is no small thing. There is an attractiveness in thinking oneself part of a group that has been preserved from errors that infect the rest of society, whether those errors are political, cultural, scientific, or religious. There is a sense of anticipation in being part of a movement that might become the wave of the future. There is satisfaction in being privy to knowledge that is unknown to the generality of mankind. It is no accident that one of the publications that repeatedly has given geocentrism room in its pages is called *The Remnant*.

Whether Protestant or Catholic, most geocentrists operate with a conspiracist mindset, though this seems to be more marked (and more remarked about) among the Catholics. Whatever their faith, geocentrists think mainstream scientists have colluded to keep the truth of geocentrism away from the public—some, because their livelihoods depend upon continued acceptance of the heliocentric model; others, because they wish to promote irreligion, the expansion of which would be retarded if geocentrism were accepted widely. It is the goal of today's geocentrists to expose this underhanded work. They do this not just through scientific arguments but through scriptural and theological ones.

Many Protestant geocentrists push the King James Only school of thought: only the Authorized Version, produced under James VI (of Scotland) and I (of England), is valid Scripture in English, and that translation should be read in a particularly literalistic way. Their Catholic counterparts are not so much concerned with a preferred translation as with literal interpretation both of the sacred text and of magisterial documents. They go so far as to say that the Church in the past infallibly taught geocentrism and still teaches it today, even if *in petto* rather than in public.

Such ideas can have unhappy consequences. People can live with bad science more easily than with bad religion. No one ever became a bad man for believing the Moon is made of green cheese, but bad men have been made from bad religion—just as good men, particularly saints, have been made from good religion. When a crabbed reading of Scripture is joined to an insupportable understanding of the physical world, today's adherent may become tomorrow's agnostic, both in science and religion. After spending time as a follower of the geocentric *gnosis*, he may conclude that sure knowledge is not attainable in either realm. The new geocentrists are keen on having people accept the truth as they understand it to be, but in the end they may leave people wondering whether truth can be ascertained at all.

The main text of *The New Geocentrists* is divided into six parts. Each part ends with a summary, "The Story Thus Far."

The first part looks at two women, Solange Hertz and Paula Haigh, who did much to legitimize geocentrism as a topic among conservative and Traditionalist Catholics. Their work, published in the 1980s and 1990s, set the stage for what came later.

The second part gives three episodes that illustrate problems with the new geocentrism and with its promoters. A young-Earth creationist attends a conference on geocentrism and leaves unimpressed. An Evangelical advises his co-religionists to steer clear of geocentrism because it can bring nothing but disappointment. An ardent geocentrist thinks he can upend Einstein's theories and advises thumbing one's nose at heliocentrists.

The third part introduces the chief Protestant proponent of geocentrism, Gerardus Bouw, who in turn introduces most of the main players in the geocentric movement. Bouw's writings are opposed by a fellow creationist who calls them misguided, and Bouw answers his critic at length. The battle is joined.

In the fourth part Bouw and two other prominent geocentrists, Robert Bennett and Robert Sungenis, are accused of plagiarism. Bennett, in particular, is shown to be a particularly sloppy borrower of other people's writings. Along the way, Sungenis's educational qualifications are called into question. If he can't represent his own background correctly, can he represent other people's ideas correctly?

The fifth part examines the temperament and arguments of Sungenis, the chief Catholic geocentrist. What do his voluminous anti-Jewish writings say about his capacity to judge fairly? How has his fascination with conspiracy theories—ranging from NASA making crop circles to the Israelis being responsible for the destruction of the Twin Towers—influence his pro-geocentrism books? He shows himself unable to handle the problem (for geocentrists) of geostationary satellites, which he confuses with GPS satellites. At a debate he leaves a listener unimpressed.

The sixth and last part takes up the people and thinking behind a new film that promotes geocentrism, even if by indirection. *The Principle* proposes that modern science is incapable of deciding whether geocentrism or heliocentrism is true. The inference is that the truth can be found only in a literalistic reading of Scripture. The Bible becomes the ultimate arbiter of scientific questions. The film is based on a misunderstanding of what happened to Earth's status when heliocentrism replaced geocentrism: its status went up, not down, contrary to the claims of proponents of geocentrism.

Throughout this book I challenge the credentials of the new geocentrists. It is fair for the reader to wonder about my own.

My undergraduate work was done at the San Diego campus of the University of California. At the time it boasted half a dozen Nobel Prize laureates and was considered one of the top three schools for mathematics in the country. I was a math major. (Later I obtained degrees in theology and law.) Students were required to take multiple courses in the sciences, including physics. They also were required to select related courses as electives.

One such elective I took, in the history of science, was a mathematical and historical investigation of geocentric theory. We used ancient observations and worked through equations and geometric analysis to see whether the Ptolemaic and Tychonian systems continued to "save the appearances" as ever more accurate measurements were accumulated. We discovered they did not.

The professor for that course was Curtis Wilson (1921–2012), then and later considered the top American expert on Johannes Kepler. It would not have been possible to take such a course from a more knowledgeable man or a more competent teacher. So impressed was I by him that I have retained his course materials for more than four decades. Some of those now-faded dittoed sheets became "How Did Kepler Discover His First Two Laws?", a fifteen-page article that appeared in the March 1972 issue of *Scientific American* and in a 1989 collection of Wilson's historical studies, *Astronomy from Kepler to Newton*.

In gratitude, I dedicate this book to the memory of Curtis Wilson.

In these pages I take the liberty of regularizing orthography. "Sun" and "Earth" are capitalized throughout, except where the latter refers not to the planet but to its constituent matter. "Aether" is preferred to "ether" to minimize thoughts of organic compounds and anesthesia. In some places I silently correct typographical errors, particularly in quotations taken from online comment boxes or blog posts.

Not so silently, I point out errors promoted by the new geocentrists. I cannot expect to convince the movement's leaders that they are mistaken, but I hope these pages may inoculate open-minded readers against a theory that should have been set aside long ago.

During the course of two decades I have had exchanges—by letter, e-mail, and online—with several geocentrists mentioned in this book. Some exchanges have been cordial, others not. Many have been frustrating. I have tried not to let my frustration or my scientific and theological disagreements color my characterization of those who are profiled in these pages. My descriptions of them are not always flattering, but I believe they are accurate.

PART 1

LADIES FIRST

I F YOU OBSERVE today's field of public geocentrists, you immediately see a commonality: they all are men. Consider the roster of speakers at the First Annual [and so far only] Catholic Conference on Geocentrism, which was held in South Bend, Indiana, in 2010 and which attracted about one hundred attendees. Advertisements listed Robert Sungenis, Robert Bennett, E. Michael Jones, Rick DeLano, John Salza, Gerardus Bouw, Martin Selbrede, Mark Wyatt, and Hugh Miller—not a woman among them. An attendee might have imagined that the modern promotion of geocentrism has been an exclusively male preoccupation, but that has not been so, at least among Catholics.

When geocentrism started to be discussed in Traditionalist Catholic publications two and three decades ago, the discussion was choreographed not by men but by women. Solange Hertz and, somewhat later, her contemporary Paula Haigh led the way. Hertz's chief writings on geocentrism appeared between 1983 and 2003, while Haigh's main work was composed at the midpoint of that span, in 1992. Only later did Catholic men begin writing with any regularity about geocentrism. Over time, the men became more prominent, and now they have the field to themselves. Largely it was a matter of age: Hertz and Haigh were born before the Great Depression, and time and infirmity removed them from active participation just as the men were finding their public footing.

I never had any contact with Solange Hertz, but beginning sometime in the 1990s I received occasional correspondence from Paula Haigh. About seven years younger than Hertz, Haigh, unlike her friend, never married. When she wrote to me, it invariably was about evolution, which she enthusiastically opposed. Some of her letters were handwritten and not always easy to decipher; other were typed. I seldom could figure out what prompted her to write to me. Occasionally she responded to something I had written but usually not. It might have been that she perceived the apostolate I founded as fundamentally orthodox (as she measured orthodoxy) except on one or two points, such as evolution or the interpretation of Genesis.

The next two chapters consider the writings of the two women. Hertz speaks of herself as a housewife and had been active in Catholic journalism in the 1970s. Haigh can claim a more structured education and displays more skill with formal argumentation. Neither has any background in science. Neither can work through the simplest equations of physics. Neither has much use of astronomical terminology. What they bring to the table is "attitude" and an insistence that the Church went off track not after Vatican II but much earlier, at least by the time of Galileo in the seventeenth century.

For Hertz, much of the allure of geocentrism arises from her posture of oppositionism. She is against many things, particularly in politics and social organization. Geocentrism fits with her pro-monarchy and anti-democracy views and with her longing for a hierarchical social order that existed centuries ago. It fits with her sense of belonging to a remnant, a core of people who have preserved the old truths and old ways against the time when the world will have need of them again. There is a romanticism about her arguments, even if they often are cluttered with offhand comments that could make even the most ardent Traditionalist Catholic wince.

For Haigh, geocentrism seems a necessary consequence of the need to protect the Bible from flaccid interpretations, and it fits with her Thomistic proclivities. To interpret scriptural passages today in a way different from that of four centuries ago—or of seventeen centuries ago, if one considers the Fathers of the Church—risks undercutting the whole of Scripture, she says. To understand a verse now in a figurative way, when it was understood in earlier times in a literal way, opens the door not only to the abandonment of biblical inerrancy but even to the abandonment of the Church itself—or so Haigh fears.

Neither Hertz nor Haigh is cited as an authority by today's geocentrists. The discussion has proceeded to a different level, one chiefly concerned with scientific claims and counter-claims, and neither woman has been much interested in the science as such. What attracts them to geocentrism is a worldview. They seem to be fairly well read and to have enjoyed the fruits of old-fashioned liberal educations. Although neither would be termed an accomplished stylist, they are competent writers. Their essays are written at a level not matched by the most prominent men, whether Catholic or Protestant, who now bear the geocentrist mantle.

THE GRANDE DAME
OF CATHOLIC GEOCENTRISM

O VER A FIFTEEN-MONTH period in 1972 and 1973, Solange Hertz wrote eight articles for *Triumph*, the magazine begun in 1966 by L. Brent Bozell, Jr., brother-in-law of William F. Buckley, Jr., who founded *National Review*. All of the articles dealt with the role of woman as wife, mother, and the center of the home. Then Hertz disappeared from the ranks of the magazine's contributors. A rift had developed regarding Vatican II. She had adopted a posture of resistance; the editors of the magazine had not.

She began writing the "Big Rock Papers," which were privately disseminated for the next decade and a half. Much of that occasional writing was used in later works. In 1974 she began publishing books that dealt with the origins of the American republic, which she said was based on Masonic principles. Everywhere in the country's history she found evidence of the errors that Pope Leo XIII, in his 1899 encyclical *Testem Benevolentiae*, called "Americanism."

She traced the problem to the Carroll family,

particularly John Carroll (1735-1815), the first bishop of Baltimore; his brother Daniel Carroll (1730-1796), one of only five men to sign both the Articles of Confederation and the Constitution; and their cousin Charles Carroll, a signer of the Declaration of Independence. The Carrolls "collaborated with the Masonic framers of the Constitution" and thus betrayed Catholicism. (Of the Declaration, Hertz said its language was such that any "true son of the Church would have held [it] under the deepest suspicion.")

In 1754, during the French and Indian War, Benjamin Franklin published his famous woodcut of a rattlesnake cut into eight pieces. The pieces represented the colonies, with the head representing the aggregated New England colonies. Beneath the image were the words "Join, or die." (It was said to be the first political cartoon in America.) Hertz saw in the image another sign of Masonry—and more than that. She saw Satan, as portrayed as a serpent in Eden. Over the years she often came back to the theme of Masonry's subversion of the American founding—sometimes this was in terms of "Judeo-Masonry"—and often she discovered marks of Satan.

(Hertz does not approve of everything written by Leo XIII. She is disappointed in his 1892 encyclical *Au Milieu des Sollicitudes*—unusual for an encyclical, it was written in French rather than in Latin—because the pontiff urged the French people, and especially the French bishops, to accept the Republic and to set aside their increasingly unrealistic hopes of monarchical restoration. He said that any form of government that promoted the common good was acceptable, at least in principle. Hertz rejects that principle.)

In the early nineties the founding editor of *The Remnant*, Walter Matt, invited Hertz to write for his biweekly Catholic newspaper. Among her early articles for *The Remnant* was one titled "The Old Religion."[1] Hertz very much opposed democracy, preferring the more ancient political form of monarchy. She saw in democracy an echo of the promise made by Satan in the Garden: "You shall be like God." In a democracy, people develop ambitions and imagine themselves being elevated beyond their proper rank; in a monarchy, society is formed in fixed gradients or classes that better reflect human nature.

Throughout human history the Evil One has been trying to undermine proper social ordering because he hates the order of heaven, which is a monarchy. This can be observed, thinks Hertz, not only in recent times but in the Middle Ages too. "The foremost Luciferian character of medieval times was certainly Robin Hood, head of a band of outlaws—one of them a transvestite called Maid Marian—who robbed the rich to give to the poor."

More on this charge about Maid Marian in a moment. First, let's continue the quotation: "As these play the heroes, the King and the poor Sheriff of Nottingham, wielding their legitimate authority under God, are cast as villains in true Cathar style. And, of course, as every English major should know, Satan is the real hero of Milton's *Paradise Lost*. After America converted the world to independence, all rebels became good, and kings and father figures bad."[2] Hertz sees political independence, in the sense of independence from the authority of a king, as an analogue to independence from Christ the King.

1. Solange Hertz, "The Old Religion," *The Remnant* (Nov. 30, 1995), 9.
2. Ibid.

Now back to Maid Marian. Hertz does not explain why she thinks Maid Marian was a transvestite, but probably it was the depiction of the legendary figure as a noblewoman with tomboyish characteristics in such films as *The Adventures of Robin Hood* (1938), starring Olivia de Haviland as Maid Marian, and *The Story of Robin Hood and His Merrie Men* (1952), with Joan Rice in that role. In such motion pictures Maid Marian at times was shown wearing trousers—at least when out riding with Robin, Friar Tuck, and the other Merry Men. Given the near-impossibility of sitting astride a galloping horse in the long dresses worn in the Middle Ages, the use of riding trousers by a woman of that era hardly seems warrant enough to label her a transvestite, a word filled with unhappy connotations.

(In the Middle Ages a woman could ride sidesaddle, but not when riding at a gallop, which is the speed at which the Merry Men had to ride when trying to evade the Sheriff. The saddles of that era did not provide sufficient control over the horse or security for the rider who sat sidesaddle. It was not until the 1830s that a new design overcame these limitations.)

The characterization of Maid Marian is not the only problem with Hertz's depiction. She says that Robin Hood was the "foremost Luciferian character" of the Middle Ages. His legend indeed goes that far back, but the story of Maid Marian's association with him does not. She first appears in the Robin Hood story only in retellings of the sixteenth century. Her figure began as a shepherdess who, with a character named Robin (but not Robin Hood), was featured in May Day pageants by the fifteenth century. It is said that this Marian—perhaps originally a personification of the Virgin Mary—over time became this Robin's love interest. Later, she was retrojected into the medieval story of Robin Hood. So it turns out that Hertz roils against a character who, even in fiction, arrived too late on the scene for Hertz's purpose.

In the same article in which she calls Maid Marian a transvestite, Hertz complains about characters in other stories. The leading figures in *Alice in Wonderland* and *The Wizard of Oz* are "gnostic," while the film *2001: A Space Odyssey* contains something more sinister. The computer HAL is a "figure of the Judeo-Christian God (Ha'El)" and "is successfully dismantled"—that is, God is dismantled, another triumph of man over his Creator.

"Also gnostic is the science fiction film *Star Wars*, whose hero, Luke Skywalker, 'a long time ago in a galaxy far, far away,' fights valiantly for 'the Rebel Alliance' against the malevolent Emperor who rules the universe": again, an analogue of man fighting against God. When the film appeared in 1977, "most critics declared the production nothing more than good, clean, noisy fun with the simplest of plots," but Hertz says that "the plot which appears so simple on the surface obviously contains subtleties grasped easily enough by adepts of the Old Religion," by which she means gnosticism. She says, "It is not hard to smell Lucifer in names like 'Luke' and 'Skywalker.'" (There is no record of Hertz commenting on the name of the author of the third Gospel.)

Three months before "The Old Religion" appeared in *The Remnant*, the newspaper published a review of Hertz's book *Beyond Politics*[3] by her contemporary and friend

3. Solange Hertz, *Beyond Politics* (Remnant Press, 1994).

Paula Haigh. That book, published by the Remnant Press, in part consists of essays that had appeared in *The Remnant*. Haigh opines that "the chapters on Judaism are of primary importance for our understanding of what is happening in the Church."[4]

Reformulating Hertz's argument, Haigh says, "Judaism is the very antithesis of Christianity in its hatred of the Cross (expiatory and redemptive suffering), in its vindictive refusal to forgive, its denial of eternal punishment, and its utopian this-worldliness." Haigh notes that Hertz identifies "the democratic process [as] Judaism's master achievement, as a religion of this world preaching equality and the brotherhood of all men, thereby realizing Judaism's deepest desires for an earthly kingdom 'whose Messiah would be perfected humanity.'" (Haigh does not have a higher opinion of Protestantism. She asks the reader, "Are you tempted—and it is a temptation—to regard Protestants as Christians?")

Beyond Politics, in Haigh's view, "outlines the frightening degree to which the conciliar Church is being Judaized." The Church is subverted through the heresy of Modernism, which was condemned by Pius X. "Modernism is the compendium of all heresies, and Judaism," says Hertz through Haigh, "is their fountainhead" (something not claimed by that pope). A result is weakness on the part of recent popes coupled, curiously, with "blind Masonic obedience to papolatry," an error in which the respect due the papacy and the scope of papal infallibility are exaggerated. (This may be the only time anyone has claimed that obedience to the papacy is Masonic.)

4. Paula Haigh, "Book Review," *The Remnant*, (Aug. 31, 1995), 8.

This suspicion of Judaism, while occasionally present, was not a hallmark of *The Remnant* of the 1990s, but over the succeeding decades the kind of argument made by Hertz and seconded by Haigh would find increasing room in the biweekly. The editors and chief writers for the publication do not themselves write in such terms, but it is telling that they publish such commentary and never remonstrate in print against the more outlandish remarks of writers such as Hertz.

In February 2003 *The Remnant* published back-to-back articles by Hertz on a quite different theme: geocentrism. The articles, under the title "The Scientific Illusion," had the now obligatory references to Freemasons, but there were no references to Jews or gnosticism. The focus was on Galileo, his purported errors, and the purported teaching of the Church on cosmology. The articles—to which we will return—came twenty years after the publication of a 7,000-word paper that Hertz titled "Recanting Galileo." This earlier paper is the core of her critique of heliocentrism. It is worth considering her comments at length before we examine the articles from 2003.

In "Recanting Galileo" Hertz says,

> there is no natural, therefore no scientific, way of knowing whether Earth is revolving around the Sun or the Sun around it. The most powerful instruments detect only movement, and this is necessarily relative. Every experiment mounted to prove heliocentricity therefore proves geocentricity equally well, depending on how data are interpreted or what the experimenter's bias is.

The Foucault pendulum, for instance, demonstrates equally well that the Earth is rotating, or that the universe is rotating around it, or that other motions are taking place. It is the same with gyroscopes, synchronous satellites, parallax, Coriolis effects, or whatever.[5]

Notice the grand sweep of her claims. Someone with a background in science would be more tentative in his phrasing, but Hertz, by her own admission, has no science training. She feels free to assert without having to back up her assertions. Some of her ideas—held in common with nearly all geocentrists—we will examine later, such as the notion that we can learn nothing regarding the truth or falsity of geocentrism from considering geostationary satellites.

Two paragraphs later, Hertz dismisses the theories for which Albert Einstein is best known: "Consequently, both the Special and General Theory of Relativity must be abandoned for cosmological purposes. Such technical studies are beyond the competence of this paper"—she should have said, "beyond my competence," since she nowhere gives any indication of actually having studied those theories—"but this should be sufficient to show that heliocentricity is still an unproved and an open question." Without offering mathematical or scientific argument, Hertz sets aside relativity and says that, with Einstein's theories now out of consideration, no one can claim that heliocentrism has been proved.

But her root interest is not the science anyway. Her

5. Solange Hertz, "Recanting Galileo," Big Rock Papers (privately printed,1983).

rejection of heliocentricity comes not from a study of celestial measurements and the crunching of numbers but from a study of Scripture. She is "fully aware of heliocentricity's potential for destroying the faith by attacking the inerrancy of Scripture" (another claim common to almost all geocentrists). In the end, "both heliocentricity and geocentricity are naturally unprovable"—that is, unprovable by natural means, such as scientific observation. "Geocentricity, however, can be proved theologically, whereas heliocentricity cannot." This for Hertz is the clincher, as it will be for later writers: science is unable to settle the question, so it is proper to defer to Scripture, which, in certain passages, must be taken in a literalistic sense that can imply only that the Earth is the motionless center of the universe.

"Thus it is clear that the structure of the universe is properly an object of faith," says Hertz. "God had to reveal that he made Earth its center, because although, like Aristotle, we might believe so from observation, on purely human faith, we could never be certain in view of the impossibility of scientific proof." Thus, "no Christian need therefore lose his sense of direction in the universe. Down is the center of the Earth, the location of hell and purgatory, to which the creed tells us our Lord descended before he arose from the dead. Up is God and his heaven."

Scientists have been wrong all these centuries, beginning with Copernicus and Galileo. "It is a matter of record that no modern scientist has yet been canonized. Even the Catholics among them, like Galileo, were not noted for their spirituality." Even if this were true, it would tell us nothing about the value of the work of particular scientists. Gregor Mendel (1822-1884) was an Augustinian friar and commonly is considered the founder of the modern study of genetics. He has not been canonized, and perhaps there is little reason to think he will be, but what does that have to do with the worth of his scientific work? (Hertz conveniently limits her comment to modern scientists, which allows her not to consider Albertus Magnus, the medieval bishop who was a precursor of modern science and who is the patron saint of scientists.)

Hertz discounts experimentation and what can be learned from it. She puts little stock in the scientific method. "With no sense of irony or contradiction, the new gnostics opted for the 'practical.' Deductive reasoning, which by its nature tends to certifiable conclusions, was abandoned in favor of inductive reasoning, which yields only to probabilities, resting as it does on constantly accumulating data." There is truth here but error also.

Hertz forgets that Aristotle—in his writings on biology, for example—worked deductively and sometimes got things wrong. He thought that thinking takes place in the heart and not in the brain, which he believed to be a cooling organ, and he believed that spontaneous generation could occur. Hertz forgets the once-accepted notion that heavier objects fall faster than do similarly-constituted lighter objects. (No, Galileo didn't disprove this by dropping weights from the Leaning Tower—that is a fable—but he did demonstrate it by rolling differently sized balls down inclined planes.) And she doesn't appreciate that "constantly accumulating data" is precisely what resulted in replacing the Ptolemaic model with the Tychonian, which itself was replaced by the Keplerian. Accumulated data, ever more precisely gathered once the telescope was invented, showed that the earlier theories no longer "saved the appearances"—but that gets us too far ahead of the story.

Just as the 2014 motion picture *The Principle* completely misstates why Giordano Bruno went to the stake, so Hertz gets the story wrong too. She says that "the renegade Dominican friar turned Calvinist" denied "the centrality of the Earth or any planet" and affirmed "the infinity of space and the existence of other worlds." Like Galileo, Bruno "was taken to task by that zealous watch-dog of orthodoxy, St. Robert Bellarmine. Duly prosecuted, he unfortunately remained obdurate and was burned at the stake in 1600." In other words, Bruno was executed because he was a heliocentrist.

It's true that he was a heliocentrist, but that wasn't why he died. Bruno was executed in Rome's Campo de' Fiori, where one can find a statue of him erected by nineteenth-century anti-clericalists. He was not tried so much for his scientific work or writings but for his pantheism and general heresy. (A minor charge against him was that he believed in a plurality of worlds, but by itself that would not have been enough to bring him to trial.) Bruno denied the Trinity, Christ's divinity, Mary's virginity, and the Real Presence of Christ in the Eucharist. He believed in the transmigration of souls—not just into other humans but into beasts.

During the twenty years before his death Bruno wandered through Europe, finding academic employment chiefly in Protestant areas but with no long-term appointments. He taught for a while at Helmstedt, where the first Protestant university had been established, but he had to flee when even the Lutherans excommunicated him.

Eventually and imprudently Bruno returned to Italy, where he was arrested and brought to Rome.

So Solange Hertz gets the story wrong. She promotes the standard secular version, that Bruno was a martyr to science, because the science he promoted—heliocentrism—she finds repugnant. After Bruno comes Galileo in Hertz's considerations. More will be said about his case later, but it is sufficient here to note that Hertz gets several things right. She points out that Galileo's "condemnation was signed by seven judges, but not by the pope, so there is no question here of an *ex cathedra* pronouncement." In this Hertz is at odds with Paula Haigh, who wrote that the condemnation of Galileo was an exercise in papal infallibility. (It was not, even if the pope had signed the condemnation, because questions of science fall outside the purview of papal infallibility.)

Hertz discounts the most famous story about Galileo: "If he ever said, 'Still, it moves,' as so often attributed to him, no contemporary recorded it." If Galileo had said *Eppur si muove*, he certainly would not have said it during his trial—even he was not that imprudent—but there is no evidence that he said it at all. The line is not mentioned in the earliest biography of him, written within fourteen years of his death by his disciple Vicenzo Viviani. The first printed report occurs in Giuseppe Baretti's 1757 work *The Italian Library*. In the same book Baretti claimed that Galileo was tortured by the Inquisition. This also is false. Not only was he not tortured (though apparently he was "shown the instruments"), but, once his trial was concluded, he was not even imprisoned. Galileo spent his last years under quite comfortable house arrest at his own estate and at other estates.

Hertz concludes "Recanting Galileo" by saying:

This is where we stand today, at a time when heliocentricity, a true heresy, has overwhelmed the entire thinking world, including the teaching organs of the Church. Rome held the line against the Enemy until the reign of Pius VII, who in 1822 [actually, 1820] finally gave limited entry to the general opinion of modern astronomers.

When Gregory XVI removed heliocentric works from the *Index* [*of Prohibited Books*] in 1835, the sequel was not hard to predict. Two modern encyclicals on Scripture, the liberal Leo XIII's *Providentissimus Deus* in 1893 and especially Pius XII's *Divino Afflante Spiritu* in 1943 (said to have been actually penned by the judaizing Cardinal [Augustin] Bea) opened the floodgates to almost any kind of "accommodated" meaning of Scripture.

This is probably the only time that Leo XIII explicitly, and Pius XII implicitly, have been called theological liberals. Be that as it may, one last point is worth noting: at the end of her long essay Hertz says that "much of the contents of this paper have been gratefully drawn" from the writings of Walter van der Kamp, the Dutch-Canadian editor of what then was called the *Bulletin of the Tychonian Society*. Van der Kamp was a Protestant, not a Catholic, and therefore was, in Hertz's eyes, a heretic. That apparently did not give her pause. She had bigger concerns than the errors of Protestantism.

Now let's take up the twin articles written by Hertz

for *The Remnant* in 2003. One appeared in the issue of February 15, and the other appeared in the next issue, dated February 28.

In the first part of "The Scientific Illusion" Hertz simultaneously offers interpretational liberality and parsimony: "Scripture does not bind us to any particular system describing the paths taken by the stars and planets as they move through space around the Earth."[6] To this point it almost seems as though she admits that Scripture does not mandate geocentrism. "We are free to espouse the Aristotelian, the Ptolemaic, the Tychonian, or any other arrangements or modifications of the many geocentric hypotheses which have been devised over the centuries." So much for the apparent liberality. "It doesn't matter whether they are concentrically organized, moving elliptically or according to epicycles, provided they seem to explain the observable movement and the Earth remains their center."

To Hertz this is not only a matter of scriptural exegesis. "Nor may it be answered that this is not a matter of faith, for if it is not a matter of faith from the point of view of the subject matter, it is on the part of the ones who have spoken." This is a confusing sentence. She says that heliocentrism is not "a matter of faith from the point of view of the subject matter." This is true, since no matter of science *per se* is a matter of faith. But she counters herself by alleging that heliocentrism indeed is a matter of faith "on the part of the ones who have spoken."

This can be understood multiple ways, and Hertz does

6. Solange Hertz, "The Scientific Illusion," *The Remnant* (Feb. 15, 2003), 12.

not explain her thinking in more detail. She might be read to mean that the authority of some who affirm geocentrism, at least implicitly, is such that the theory is elevated to a matter of faith. She appeals to the most prominent character in the Galileo episode other than the scientist himself, Cardinal Robert Bellarmine (1542-1621), and by extension to the writings of the ancient Fathers of the Church.

She quotes the letter written by Bellarmine to Paolo Antonio Foscarini (c. 1565-1616), a Carmelite priest and scientist whose book on the mobility of the Earth was condemned at the same time as Galileo's writings. Bellarmine reminded Foscarini that the Council of Trent, which had concluded in 1563, "prohibited expounding the Scriptures contrary to the common agreement of the holy Fathers. And if Your Reverence would read not only the Fathers but also the commentaries of the modern writers on Genesis, Psalms, Ecclesiastes, and Joshua, you would find that all agree in explaining literally that the Sun is in the heavens and moves swiftly around the Earth, and that the Earth is far from the heavens and stands immobile in the center of the universe."

We will come back to these several points later. The role of Bellarmine in the controversy is stressed by modern proponents of geocentrism, and the Catholic proponents in particular argue that early Christian writers unanimously endorsed geocentrism and that this was tantamount to an infallible teaching on the subject.

Galileo proposed his theories after Copernicus had proposed his. Galileo was brought to trial, but Copernicus was not. The latter even had his work praised by a cardinal, Nicolaus Schoenberg of Capua, who in 1536 encouraged

Copernicus to "communicate your discovery to the learned world." Why the different treatment? As Hertz puts it, "until Galileo came along, no one dared seriously pretend that [heliocentrism] reflected reality." That is a loaded sentence, with "dared seriously pretend." What actually had been happening is that it was generally agreed that no theory of the movement of the heavenly bodies, neither heliocentrism nor any variant of geocentrism, "reflected reality." In the absence of later-invented scientific instruments, there was no way to prove one theory over another.

It was expected that an astronomer would lay out a theory, showing how well it "saved the appearances," without asserting that the theory accurately illustrated what one would see if one were positioned a vast distance from the visible cosmos. This is where Galileo overstepped. He was unable to prove his theory—he knew he could not, even with the help of the recently-invented but still modest telescope—but he wrote as though heliocentrism not just accommodated observations but explained them in actuality.

(In what I consider to be his most interesting book, *The Discarded Image*, C. S. Lewis made this observation about the Galileo case: "The real reason why Copernicus raised no ripple and Galileo raised a storm may well be that whereas the one offered a new supposal about celestial motions, the other insisted on treating this supposal as fact. If so, the real revolution consisted not in a new theory of the heavens but in 'a new theory on the nature of theory.'"[7])

Hertz claims that until the time of Galileo, "the whole

7. C. S. Lewis, *The Discarded Image* (Cambridge: Cambridge University Press, 1970 [1964]), 16, quoting Owen Barfield, *Saving the Appearances* (London: Faber and Faber, 1957), 51.

world, both Christian and pagan, believed that our Earth was the center of the universe and that the entire cosmos revolved around it, *because geocentricity is a truth of the natural order revealed by God from Adamic times* [her italics]. Inasmuch as it's impossible to see what's actually going on in outer space without standing outside the universe, this truth is a proper object of revelation." She seems to mean that, if we can't know an answer through science, it will be present to us through theology. This does not follow.

With respect to geocentrism, we *can* know the answer through science—at least enough to demonstrate that geocentrism can't be true—but, even if we could not know the answer scientifically, that would not imply that the answer must be generated for us through theology. There are other problems that have no definitive scientific solution—such as Heisenberg's uncertainty principle: that the more precisely we know one variable, such as a particle's location, the less precisely we know another variable, such as the particle's momentum—and yet no one claims their solutions are to be found in the pages of Holy Writ.

Hertz characterizes Galileo's chief fault this way: "according to him, scientific certainty was attainable by the natural sciences on their own without the help of revelation." Galileo "believed the obvious sense of Scripture must give way before scientific evidence to the contrary. In other words, he maintained that because the Bible's competence extended only to spiritual matters, it could, and in fact does, contain many errors in the natural order and was not to be trusted in the scientific world of hard facts."

More properly put, Galileo's fault was that he arrogated the interpretation of Scripture to himself. It was not that he expounded the Copernican theory—after all, Copernicus himself did the same and was not brought up on charges before the Inquisition—but that he presumed to say how Scripture should be interpreted. He compounded his problems by presenting his thoughts in ways guaranteed to annoy those in authority, including even the pope.

Solange Hertz—without formal training in either science or theology—ends the first of the pair of articles in *The Remnant* with the claim that "heliocentrism remains proscribed today and may not be knowingly and willfully entertained or promoted by any Catholic without endangering his salvation." In her follow-on article, the second part of "The Scientific Illusion,"[8] Hertz complains about John Paul II's "rehabilitation" of Galileo during an address he gave to the Pontifical Academy of Sciences in 1992. The secular media, not surprisingly, spun the Pope's words to the Church's disadvantage. Hertz quotes the report in *The Washington Post* as saying, "The Pope noted that Galileo rejected the Church's suggestion that he present the Copernican system as a hypothesis, instead of demonstrated truth. No one at that time had laid out 'irrefutable proof' of the Copernican model, the pontiff said."

Hertz replies: "Alas, the article neglects to mention that this irrefutable proof has yet to be supplied by anyone, and that even atheist scientists are beginning to abandon heliocentrism, along with Newtonian physics." She does not say where she got the notion that "atheist scientists" are leaving heliocentrism for geocentrism. She gives no names and no source for her information. She may have been the

8. Solange Hertz, "The Scientific Illusion," *The Remnant* (Feb. 28, 2003), 9.

first Catholic writer to make this claim. She would not be the last. In the following years other proponents of geocentrism would say that many unbelieving scientists were coming over to geocentrism, but no corroboration would be offered.

Hertz is unable to conclude her article without bringing in the diabolical and Masonry:

> Satan and his think tanks inspired their tool Galileo to reverse the order of the universe in men's minds, persuading them that with the Sun at the center, the Earth had no special importance. . . . When science declared its independence from the faith which till then had preserved it from serious error, scientists "professing themselves to be wise," as St. Paul says, "became fools" (Rom. 1:22) who, by tailoring the faith to their reasoning, ended up abandoning it altogether. . . .
>
> Freemason-alchemist Isaac Newton began constructing a whole new universe based on a mathematical system existing largely in the human mind, where matter literally moved itself by so-called "laws of gravity," which in due time were generally accepted as true with no more proof behind them than Galileo's heliocentrism.[9]

Several things can be said about this. First, any "mathematical system" exists "largely in the human mind." After

9. Ibid., 12.

all, mathematics is a purely abstract science. This is as true for Newton's mathematics as for the mathematics used by Ptolemy and Tycho Brahe. Second, matter is not moved "by so-called 'laws of gravity'" but by gravity. The laws of gravity merely seek to put into formulas what is observed in nature. The laws of gravity can be tweaked over the centuries—and they have been, as gravity has become better understood—without discounting the reality of gravity. If gravity or the laws that describe its workings are not worthy of respect—if they don't have any "more proof behind them than Galileo's heliocentrism"—then how can one avoid saying the same about any scientific theory or discovery?

(Over the last two centuries scientists such as James Clerk Maxwell developed laws of electromagnetism that allowed the invention of devices that Solange Hertz has in her home and uses while writing, such as electric lights. She nowhere is on record as referring to Maxwell's laws as the "so-called 'laws of electromagnetism.'")

As almost an anticipatory critique of these articles, in late 2000 a priest of the Old Catholic Church, Charles T. Brusca, wrote a letter to *The Remnant* complaining about an article written by Solange Hertz. That article appeared in the November 15 issue. She claimed, according to Brusca, that "the theories of Newton and Einstein contradict the Catholic faith." *The Remnant* didn't print Brusca's letter, perhaps because Brusca belonged to a schismatic church, so he reproduced it in his parish bulletin the following February.[10] His comments could have applied equally well

10. Charles T. Brusca, unpublished letter submitted to *The Remnant*, (Jan. 16, 2001), rosarychurch.net/answers/qa022001b.html.

two years later regarding Hertz's two-part series on "The Scientific Illusion."

In his bulletin Brusca prefaces his letter by saying that Hertz's 2000 article was "especially endorsed by [*The Remnant's*] editor," who by that time was Michael Matt, the son of the newspaper's founder. The article "contains a number of foolish notions," says Brusca. "Regular readers know that *The Remnant* often contains articles about 'what should have been' rather than about 'what actually happened' or 'what can now be done.' Usually they restrict themselves to political commentary on how 'every government should be a Catholic monarchy' or on how 'everything will be okay if we can just have the indult Mass.' The article in question is a bit more obviously confused and would be an embarrassment to Catholics who think that *The Remnant* consistently represents the Church's tradition—which it often does not."

Brusca labels "disturbing" "Hertz's denial of the physical realities around her in the name of traditional Catholicism. . . . Newton's explanation of gravity is incomplete, as was Aristotle's and as Einstein's is likely to prove to be. But that does not make [gravity] any less real! . . . Hertz's claim that Newton's and Einstein's theories were blindly accepted without proof suggests that she has not bothered to read anything about the theories and their history. Precisely the key note of modern science, and the thing that made it so unsettling to seventeenth-century society, was that it turned toward experimental verification and away from 'authority.' Instead of simply accepting what Aristotle had to say about the movements of bodies, people began to measure the movements for themselves."

As for Cardinal Bellarmine, he "was insistent in not

overthrowing the teachings of the Fathers in biblical matters concerning the natural sciences—until there was compelling proof that the Fathers had been hampered in their interpretation by inaccurate information. He had simply not been shown anything compelling." In this Brusca the non-Catholic understood better than Hertz the Catholic. Bellarmine's position was that one must adhere to the ancient cosmology in the absence of compelling proof of any then-modern variant. If such proof were forthcoming, he would have accepted it and sought for interpretations of Scripture that accommodated it, there being multiple interpretations possible for many passages.

Hertz will have none of that. To her, Bellarmine was unmovable in his adherence to geocentrism (though that is not what he said), and she rejects what she considers to be liberalizing trends in scriptural interpretation as taught by modern popes. Brusca had the foresight to see that his recommending to Hertz that she read *Providentissimus Deus* would be fruitless: "her low regard for Pope Leo XIII would probably make that a waste of time."

Brusca closes his letter by saying that "*The Remnant* is getting to be more and more of an embarrassment to traditional Catholicism as time goes on, and it becomes increasingly clear that it is a journal devoted to seventeenth-century antiquarian thinking." He shakes his head over "the wistful desire for monarchy and now the denial of physical reality."

CONVINCING BUT FALSE

S INCE LATE 2012, while residing in a retirement home, Paula Haigh has had a blog titled "Catholic Creation Cosmology." She admits having no computer skills. A friend has to transfer her writings to, and has to maintain, the blog. The opening lines of her "Mission Statement for This Blog," uploaded in November 2012, read this way: "Of all the fruits of faith and reason—or one should say, not only of reason but also of faith—that of Creation is absolutely primary. Presupposing God's existence, a truth both of reason and of faith, the very first article of the earliest creeds is that God is creator of heaven and Earth (*creatorem coeli et terrae*)."[11]

So far so good. Then Haigh turns to her private interpretation of the opening chapters of Genesis, which she takes every bit as literally as any Protestant Fundamentalist, but she immediately parts company with most Fundamentalists

11. Paula Haigh, "Mission Statement for This Blog" (Nov. 22, 2012), coelietterrae.blogspot.com/2012/11/the-importance-of-creation-and-church.html.

because she rejects heliocentrism, which, she believes, goes hand in hand with her *bête noire*, evolution.

She says, "These facts have been thoroughly documented most recently by Robert Sungenis in his monumental *Galileo Was Wrong: The Church Was Right*, two volumes of true science and true history, exposing the Luciferian lie of evolutionary modernism as defined by *Pascendi*, the landmark encyclical of 1907 by Pope St. Pius X, and the infallible ruling of Urban VIII in 1633." We will turn to Sungenis in later chapters, but it is worth noting here three things: (a) his book on geocentrism, which originally was published in two volumes, now appears in three; (b) *Pascendi* contains no reference to geocentrism; and (c) the ruling issued by the Holy Office and approved *in forma communi* by Urban VIII in 1633 is not considered by theologians or Church historians to have been an exercise of papal infallibility.

Near the end of her introductory blog post, Haigh says that "a literal six-day creation week, including a geocentric cosmology today, may call forth only ridicule and even persecution. But truth is truth." She believes that she knows the truth. In a post written a month later, in a consideration of Robert J. Spitzer's *New Proofs for the Existence of God*, Haigh writes, "It must be said also that I am not a scientist as Fr. Spitzer is. Fr. Spitzer has advanced degrees in the physical sciences, whereas my A.B.D. is in literary theory. I just happen, by God's providence, to have had the immense privilege of studying under three men who were devoted students, not only of their specialties, but also of St. Thomas Aquinas."

Haigh's writing is suffused with quotations from, and

citations to, the Angelic Doctor. It is worth keeping in mind that she has no science background whatsoever, and even her studies of Aquinas have been chiefly on her own; her academic area was not theology but literary theory.

Four months after publishing her introductory post, Haigh took to task John Vennari, editor of *Catholic Family News*, a monthly newspaper that may be the second-most influential Catholic Traditionalist publication after *The Remnant*. The publisher of *Catholic Family News* is Nicholas Gruner, a priest best known for promoting an idiosyncratic view of the Fatima apparition. Fatima is as important to Gruner as anti-evolutionism is to Haigh: each has discovered a key that opens many doors. Gruner was ordained in the diocese of Avellino, Italy, in 1976. In 1996 the bishop of that diocese suspended Gruner's priestly faculties. Gruner appealed the suspension. The case went to the top of the Church's legal system, the Apostolic Signatura in Rome, which confirmed the suspension in 2001. Gruner claims never to have been suspended at all.

In Haigh's opinion, Vennari "and all other Traditionalist leaders and editors" (presumably including Michael Matt of *The Remnant*) "persist in your failure to see—or is it a willing blindness?—that the Second Vatican Council's 'original sin' is that of believing that the Church, like all other institutions on planet Earth, must evolve—that is, change or go extinct! This is made clear in paragraph 26 of *Pascendi*. Archbishop [Marcel] Lefebvre [founder of the Society of St. Pius X and a hero to many Traditionalists, including Vennari] not only lacked Catholic vision but was unable or refused to learn from the truly Catholic vision of Pope St. Pius X, who is the one who has warned the Church of the evolutionary, Modernist threat. It is an evolutionary ideology that permeates the documents of Vatican II, and those who refuse to see it, and point it out, will surely have much to answer for."

Today Haigh seems to be quite estranged from the wider Traditionalist movement, but it was not always so. Though she never contributed as widely to the movement's publications as did her friend Solange Hertz, she did submit regular letters to the editor and the occasional book review—for example, a review, mentioned in the previous chapter, of Hertz's *Beyond Politics*.

If Haigh's opinions have not evolved over the last few decades, perhaps her sensibilities have. Once she held back from criticizing other Traditionalists, but no longer. To her mind, most Traditionalist writers and leaders have let down their own movement because they have failed to speak against Vatican II's supposed roots in evolutionary theory. They have opposed the council chiefly on less fundamental grounds. For Haigh, evolution is the root of all error; it can be traced even as far back as Galileo, though of course the term was not then in use. If many Traditionalists consider Marcel Lefebvre (1905-1991) their hero, certainly Haigh's is Pius X (reigned 1903-1914).

Given that pope's focus on Modernism—the adherents of which said that doctrine can evolve, not just develop—one might expect that Haigh's most voluminous writings would be on that heresy or on biological evolution, but not so. Her most concentrated work has been on the Galileo affair and its consequences. In 1992 she wrote, and in 1999 revised, a series of three long essays: "Galileo's Heresy," "Galileo's Empiricism—and Beyond," and "Was It/

Is It Infallible?" They total 56,000 words, the equivalent of a 200-page book. Let's look at them in order.

After referring to a French writer whose works have not been translated into English, Haigh begins "Galileo's Heresy"[12] by praising Solange Hertz's essay "Recanting Galileo." She says that Hertz's writing "always possesses a spiritual dimension not to be found anywhere else." Perhaps, but the most obvious difference between Hertz's writing and Haigh's is that the latter is much more philosophical, attempts to be more precise in reasoning, and relies less on airy historical and political statements. Haigh seems to be an almost-scholar who has maintained a lifelong passion for the works of Aquinas and who understands the need to repair to the library for research. Hertz happily refers to herself as a housewife; she freely shares her opinions, which are phrased less formally than Haigh's and often more provocatively.

After Haigh's praise of Hertz comes mention of Walter van der Kamp (1913-1998), a Dutch Protestant who settled in Canada and founded the Tychonian Society "and its quarterly journal, *The Biblical Astronomer*, formerly known as *The Bulletin of the Tychonian Society*." When John Paul II was contemplating Galileo's "rehabilitation," van der Kamp delivered a letter to him, urging him to reconsider. The Pope went ahead anyway.

Next Haigh mentions Gerardus Bouw, van der Kamp's successor and also a Protestant, but the praise is conditional: "One must beware, however, of Dr. Bouw's very

12. Paula Haigh, "Galileo's Heresy" (1992; revised ed. 1999), ldolphin.org/geocentricity/Haigh2.pdf.

anti-Catholic prejudices which sometimes cause him to distort history." We will look at Bouw's work in subsequent chapters. Haigh ends her list of recommended writers with the name of Marshall Hall, author of *The Earth Is Not Moving*, "a quintessentially popular treatment of this difficult subject, and he must be given much credit for bringing the arcana of modern mathematical physics down to the level of us scientifically illiterate mortals."

In describing herself as such—the characterization could apply also to Hertz—Haigh inadvertently brings focus to a curious fact: most proponents of geocentrism have no or very little training in science. Some seem never to have taken a course in physics, and many of the most public of them, when referring to experiments or theories, give no indication that they actually understand the underlying mathematics or the physics described by the mathematics.

After quoting Martin Gardner (who for twenty-five years wrote the "Mathematical Games" column for *Scientific American*) concerning frames of reference, Haigh draws the conclusion that "there is simply no human way of knowing the structure of the universe. But God has revealed it! This was the basis on which Galileo was condemned by the Holy Office in 1633. It is, therefore, a fact of divine revelation, a truth of faith." This, basically, is her whole argument. Incapable of grappling with the science as such, she falls back on the idea that the Holy Office's condemnation of Galileo amounted to an exercise of the Church's infallible teaching authority.

This is premised on her understanding that "his heresy was specifically to doubt the inerrancy of Holy Scripture," even though in his "Letter to the Grand Duchess Christina" (1615) Galileo argued, in Haigh's words, that "the Scriptures are not to be interpreted literally when they speak of physical things but only when they teach on matters of faith and morals." (This isn't quite accurate, but we'll let it pass.) "Galileo well knew that the Fathers of the Church held to a geocentric view of the universe and taught the same in a unanimous way as any other view would have been immediately recognized by them as against Scripture and common sense or reason. . . . Galileo shines forth as the first Modernist, for he distorts the Sacred Scriptures to fit his own opinions."

It is not easy to demonstrate that many of the Fathers ever wrote about cosmology at all; it is not easy to show a "unanimous consent" when most of them apparently had nothing to say on the topic. Like everyone else in ancient times, when they wrote about the heavens, they wrote in terms of appearances (the Sun rises and sets; the stars move through the sky), without any attempt to formulate astronomical theories.

In Haigh's opinion, "The real centerpiece of the Galileo affair is the letter that St. Robert Bellarmine wrote to the Carmelite friar Paolo Antonio Foscarini after reading Galileo's letter to Castelli and Foscarini's sixty-four page book defending the compatibility of the new Copernican system with Holy Scripture." (Benedetto Castelli, a Benedictine priest, was professor of Mathematics at the University of Pisa. The letter to him was the draft of what would become Galileo's "Letter to the Grand Duchess Christina.") Haigh says, "Cardinal Bellarmine assures us that the consent of the Fathers and their commentators is unanimous in holding a geocentric and geostatic view of the universe based on Holy Scripture. . . . Galileo and the heliocentrists or Copernicans attacked a truth of faith, namely, that Holy Scripture is inspired and inerrant in all its parts and that we may not depart from the common agreement of the Fathers in our interpretations."

Haigh assumes much here:

> That Fathers of the Church intentionally taught cosmology.
> That there was "unanimous consent" among them regarding the topic.
> That this "unanimous consent" elevated this matter of science to the level of doctrine.
> To deny the geocentric view implies that one denies the Bible's inspiration and inerrancy.
> That Bellarmine's evaluation of the matter itself somehow is definitive, even if not infallible.

The crux of Haigh's argument is that "the Church's decisions and pronouncements in the Galileo case were indeed infallible." She acknowledges that

> It may be objected that based on the Decree of Infallibility from Vatican I—which declares papal infallibility only when [the pope] "defines a doctrine regarding faith or morals to be held by the universal Church"—the Church could not or did not pronounce definitively or infallibly upon a teaching of Holy Scripture that concerned matters of physical science.

However, it seems to me that this is precisely what we are to learn from the Galileo case—that the Church, by reason of her appointment as supreme guardian and interpreter of Holy Scripture and the entire deposit of faith, can and must tell us the true meaning of Scripture, and this infallibly, whether the Scripture speaks in that instance of natural or of supernatural things.

Notice the telling words: "it seems to me." It seems to Paula Haigh that the Church interprets passages of Scripture for us, even those apparently dealing only with natural phenomena—what later would be termed science. But this is not what Vatican I taught, and it is not how popes have taught. The Church has not been in the habit of issuing definitive interpretations of lines of Scripture. Only a minuscule number of passages have been so interpreted. The Church leaves interpretation mainly in the hands of exegetes who do not themselves possess the note of infallibility—even the sainted ones, such as Robert Bellarmine.

Haigh concludes this first of her trio of essays by noting that "Galileo's sentence was commuted; his daughter, a Carmelite nun, was allowed to recite the penitential psalms in his stead [part of his penalty was to recite those seven psalms weekly for three years], and Galileo passed the remaining years of his life on his own country estate at Arcetri, near Florence, working on his final and probably favorite work, *Discourses on Two New Sciences*, the work which has earned him the title of the Father of Modern Physics." Galileo was let off easy, but "the decree of the Holy Office against Galileo has never been abrogated—nor can it be. The wording is quite absolute." Haigh insists that the decree, not having been abrogated (repealed), must still be in force not only as a legal document but as an interpretive document.

A decade before Haigh wrote these words, John Paul II acknowledged that errors were committed by the tribunal that judged Galileo's scientific writings. The Pope's rehabilitation of Galileo came in consequence of a report issued by a committee of the Pontifical Academy of Sciences. True, the Pope did not abrogate the Holy Office's 1633 decree against Galileo; the decree was not erroneous in all its parts, nor were the penalties imposed necessarily unreasonable, especially considering the times. But the Pope noted that Galileo had the underlying science basically correct, even if he overstepped his bounds in trying to interpret Scripture.

At 29,000 words, Haigh's second essay is the longest of the three. "Galileo's Empiricism—and Beyond"[13] is not so much about geocentrism as about the scientific method and about the development of Newton's laws of motion and especially his ideas concerning gravity. That is the force, thought Newton, in terms of which we can explain the motions of the objects we see in the sky.

Haigh disagrees. She subscribes to the idea that the planets and other celestial bodies are moved through the superintendence of angels, and she is wary of those who think otherwise—especially those who lived a long time ago. In a discussion of a book by Stanley Jaki, the priest-physicist, Haigh reproduces a quotation he provides from

13. Paula Haigh, "Galileo's Empiricism—and Beyond" (1992; revised ed. 1999), ldolphin.org/geocentricity/Haigh.pdf.

John Buridan, a fourteenth-century priest whose work was a precursor to Copernicus's. Buridan wrote:

> Since the Bible does not state that appropriate intelligences move the celestial bodies, it could be said that it does not appear necessary to posit intelligences of this kind, because it would be answered that God, when he created the world, moved each of the celestial orbs as he pleased, and in moving them he impressed in them impetuses which moved them without his having to move them any more except by the method of general influence whereby he concurs as a co-agent in all things which take place. . . .
>
> And these impetuses which he impressed in the celestial bodies were not decreased nor corrupted afterwards, because there was no inclination of the celestial bodies for other movements. Nor was there resistance which could be corruptive or repressive of that impetus.

"Of this passage," says Haigh, "Fr. Jaki says that Buridan's statements 'anticipate Newton's first law of motion.' We grant that they do, and we go farther than Fr. Jaki and protest that they do so by removing the angels as movers of the celestial bodies." While Haigh certainly isn't alone among modern geocentrists in ascribing movements of planets and stars to angels, she likely is in the minority—at least if one presumes that silence on the issue by other geocentrists indicates disagreement with her position. If she agrees with her friend Solange Hertz—"Down is the center of the Earth, the location of hell and purgatory, to which the creed tells us our Lord descended before he arose from the dead. Up is God and his heaven"—then there is a problem that neither she nor Hertz addresses.

In the old cosmology, as used by Dante in his *Divine Comedy*, the worst place—hell—is located at the center of the Earth, and the best place—heaven—is at the furthest remove from the Earth, beyond the sphere of the stars. The further from Earth, the more spiritual and holy; the closer to Earth, the more material and corruptible. In this framework, it could make sense to speak of angels moving the stars because those heavenly bodies are close to heaven, the abode of angels.

But what about the Moon? It is the heavenly body closest to Earth and thus closest to the realm of corruption. Would angels be assigned to move it—or would demons? If the latter, where in outer space is the divide at which demonic movers give way to angelic movers? If angels move even the Moon, who moves things still closer to Earth, such as meteoroids that pass close by or meteorites that strike the Earth? One could pursue such an inquiry indefinitely, ending with having an angel assigned to anything of any size that moves anywhere—including even Newton's fabled falling apple. Haigh gives no particular defense of angels as movers of physical objects great and small; she seems to accept the ancient notion as fact rather than as a best-guess developed long before the physical sciences as such came to be.

Haigh worries that Newton's laws of motion remove God from the cosmos. "Newton's second law is even more explicit in its reference to impetus or impressed force, but there is not a hint that God could be involved in any way in this 'motive force.'" She is unwilling to entertain the idea (almost universally accepted by believers for centuries now) that God remains "involved" by (a) keeping things in existence at all and (b) endowing his creation with forces such as gravity that can account for movements.

Haigh quotes Herbert Thomas Schwartz, who wrote an essay on "The Five Demonstrations of the Existence of God" for Benziger Brothers' 1948 publication of the *Summa Theologiae*: "Where St. Thomas says that nothing is moved unless it is moved by another, Newton says that motion continues unless it is stopped by something else." Schwartz says these notions "could not be more sharply opposed," but that is not the case at all, if God, the Unmoved Mover, chose to effect motion through the institution of gravity and other forces.

But none of this would make an impress on Haigh. She gives lengthy quotations from Aquinas and then concludes: "These quotations should make abundantly clear, by contrast, the damage done to men's souls by the ruthless excision of God from descriptions of motion." She might as well complain about the "ruthless excision" of God from descriptions of plowing or highway construction. Despite the extended quotations she serves up from Aquinas, Augustine, and other saints, she is determined to maintain a cosmological worldview that they likely would have dropped had they lived in much later times. She is obdurate: "Fundamentalism, anyone? Make mine Catholic."

She ought to observe the advice given by Augustine in a quotation that she provides: "We should always observe that restraint that is proper to a devout and serious person and on an obscure question entertain no rash belief. Otherwise, if the evidence later reveals the explanation, we are likely to despise it because of our attachment to our error." Indeed. If the error is attributing celestial motion not to gravity or mechanical forces but to the shoulders of angels, so to speak, then it is no virtue to remain attached to that error. If later evidence has revealed a sufficient explanation, such as in Newton's laws, then accepting such an explanation no longer can be considered rash, even if it legitimately might have been considered rash before or while the explanation was under development.

Haigh sums up her thoughts with a quotation "from the work of a fellow Catholic who puts so well what we have lost by the error of heliocentrism. James Foresee in his *The Heliocentric Hoax* writes":

Before the Galileo heresy the Christian, as opposed to the progressive modern man, was not only geo-centric but theocentric (God-centered). Before the "Earth-movers" arrived on the scene, Western civilization had an orderly world view; everything had its place. . . . All was orderly and secure; man believed, and he belonged. Then, with the new world view, came doubt, the enemy of faith. . . .

Man, now displaced from the center of the uni-verse, not only sustained a loss of dignity, purpose, direction—but also he was most tragically, psy-chologically divorced from God the all-unifying Creator. That is why this controversy is crucial.

Those are sentiments accepted not just by Paula Haigh but by most modern geocentrists. Man has lost a sense of "dignity, purpose, direction." With the dissolving of the old world view, man finds himself stranded. He no longer knows his place, either with respect to God or with respect to his neighbors. He used to feel important as he surveyed the skies. He understood that everything revolved around him. Now he feels unimportant. (Perhaps he should feel humbled instead; perhaps he chafes with injured pride.)

The promotional material for the pro-geocentric motion picture *The Principle* asks, "Are you significant?" The implication is that you can't be significant if the planet you inhabit is not in a significant location, and there is only one location that truly can be called significant: the very center of the universe. Nothing else counts. If the Earth is located anywhere else—if it is located on the fringe of one arm of an undistinguished spiral galaxy lost among countless other galaxies—then you can have no real significance. Your significance arises not so much from the fact that God made you, loves you, and superintends you but from your spatial location. Your worth is a function of your celestial street address.

By the logic used by Haigh and others, if Earth is significant only because it is located at the center of the universe, then more significant still must be the very center of the Earth, because the center is ultimate physical centrality. Yet the center of the Earth is where—according to such people—hell is found. Conversely, by their understanding the least significant place must be the point furthest from the central Earth, a point beyond the sphere of the stars, which is where heaven is said to be found. (This brings a new meaning to William Butler Yeats' most famous line: "Things fall apart; the centre cannot hold.")

Rounding out Haigh's trio of essays is "Was It/Is It Infallible?"[14] This essay is half the length of the previous one, but it contains more material pertinent to this book. Haigh begins with a look at two authors writing more than a century ago. Andrew Dickson White, co-founder and first president of Cornell University, was the author of *A History of the Warfare of Science with Theology in Christendom*.[15] He wrote against what he considered to be religious obscurantism. He thought religion should not interfere with science and really ought to be subservient to it.

White was opposed by James J. Walsh, who taught the history of medicine at Fordham University. He is best known for *The Thirteenth, Greatest of Centuries*,[16] but another book of his was popular a hundred years ago, *The Popes and Science*.[17] "This latter book is a running polemic" with White's book, says Haigh, who complains that "Walsh, the Catholic, was not able to see the supernatural dimensions of his subject . . . and was intent upon defending the popes and the Church in the natural order only, as the popes were patrons of scientific progress in empirical and experimental methods and results." In Haigh's opinion

14. Paula Haigh, "Was It/Is It Infallible?" (1992; revised ed. 1999), ldolphin.org/geocentricity/Haigh3.pdf.
15. Andrew Dickson White, *A History of the Warfare of Science with Theology in Christendom* (New York: George Braziller, 1937 [1895]).
16. James J. Walsh, *The Thirteenth, Greatest of Centuries* (New York: Catholic Summer School Press, 1920).
17. James J. Walsh, *The Popes and Science* (New York: Fordham University Press, 1908).

Walsh, like White, "accepted the Copernican error as truth." Walsh wrote:

> It was rather because of the way in which Galileo urged his truths than because of the truths themselves that he was condemned. Even Professor [Thomas] Huxley, in a letter to Professor St. George Mivart, November 12, 1885, said: "I gave some attention to the case of Galileo when I was in Italy, and I arrived at the conclusion that the pope and the college of cardinals had rather the best of it."

"Huxley," says Haigh, "was referring here not to Galileo's pugnacious and arrogant attitude but to the fact that he could produce nothing remotely resembling the proofs and demonstrations that Cardinal Bellarmine had required before he would consider the Scriptures and the Fathers seriously challenged by the Copernican astronomy."

This comment by Haigh is more telling than she realizes. She is correct to note that, given the procedural format of Galileo's trial, it was incumbent upon him to provide sufficient "proofs and demonstrations," but he could offer no scientific evidence stronger than that used in favor of geocentrism. Although he made use of the telescope, Galileo's device was far simpler than telescopes that would be developed in later years, and he was unable to observe such things as stellar parallax. He had observations and arguments drawn from his observations, but he did not have enough evidence to dethrone geocentrism.

So to this extent Haigh is correct. She also is correct in a way that she may not see. By noting that Bellarmine was looking for "proofs and demonstrations," she implicitly acknowledges that, had such things been produced by Galileo, Bellarmine would have agreed that the then-current understandings of some passages of Scripture were being "seriously challenged." The implication is that Bellarmine was open to the possibility of such a challenge being mounted in the future—he just did not see Galileo mounting anything close to it before or during the trial.

Haigh turns to an argument that many Catholic geocentrists think is the clincher: the unanimous consent of the Fathers of the Church. She says, "The geocentric and geostatic cosmology is based on Holy Scripture and is endorsed by the unanimity of the Fathers—an infallible sign of truth." She quotes from *The Channels of Revelation*,[18] written by Emmanuel Doronzo (1903-1976), who taught sacramental theology at Catholic University of America:

The doctrine of the Fathers has a particular value and a greater force in any argument drawn from Tradition, because it is scientifically the surest and easiest way to find the truth; this is the reason why the argument from Tradition is often called "the argument from the Fathers" and is confined mainly to their doctrine.

In the making of such an argument, the texts should be critically and certainly established, the doctrine should be carefully evaluated to make sure

18. Emmanuel Doronzo, *The Channels of Revelation* (South Bend: Notre Dame Institute Press, 1974).

that the Fathers speak of a doctrine about faith or morals and propose it positively (not merely opinionatively) as something to be held with faith, and finally the morally unanimous agreement of the Fathers on such a doctrine should be established.

Haigh says, "With regard to the phrase above concerning 'faith or morals,' we know that heliocentrism and/or a-centrism [the idea that there is no center or that nothing is at the center of the universe] are against faith because the condemnations on which Galileo's heresy are based tell us that the doctrine that the Sun does not move is 'formally heretical' and the doctrine that the Earth moves is 'at least erroneous in the faith.' . . . To hold, therefore, that the geocentric/geostatic cosmology has nothing to do with our Catholic faith is to cut us off from infallible authentic Tradition." She continues with a long quotation from Doronzo in which the theologian lists multiple commendations of Thomas Aquinas from Leo XIII through Pius XII. No commendation alludes to cosmology.

Then Haigh concludes this section of her essay: "And so, it is certain from all the above that a geocentric/geostatic universe was held by all the Fathers; and we know that it was also the view of St. Thomas Aquinas, the Church's greatest theologian. Therefore, the geocentric/geostatic essence of the medieval cosmology is infallible truth and *de fide*. It is an integral part of the Deposit of Faith and of apostolic tradition."

Fine—except that she is wrong. If Galileo failed to prove his case when brought before the Inquisition, here Haigh fails to prove her case. Her most basic error is thinking that the Fathers of the Church taught geocentrism as they taught Christological or sacramental doctrines, as beliefs to be held as matters of Catholic faith. Did they accept geocentric cosmology? Yes, as did probably everyone at the time. There was no unreason in that. A geocentric view is the default position of people living in an era prior to the development of the physical sciences and scientific instruments. But did the Fathers *teach* geocentrism as part of the faith? There is no evidence for that.

It is sufficient to note that the Fathers accepted a geocentric cosmology as they accepted the notion that the world is constructed of four constituent elements (air, earth, fire, water) and that bodily illness is a consequence of imbalances among the four humors (black bile, yellow bile, phlegm, and blood), each of which corresponds to one of the four temperaments (melancholic, choleric, phlegmatic, sanguine).

The Fathers, like others of their eras, accepted the four elements and the four humors as adequate explanations of what they saw around them. Geology was then a nonexistent science, and medicine was rudimentary. The term "geology" was first used in 1603. Bacteria were unknown until discovered by Antonie van Leeuwenhoek in 1676, and aspirin—the world's first effective analgesic—was not developed until the end of the nineteenth century. The Fathers of the Church likely would have set aside the notion of the four elements if they had had at their disposal even the geologic knowledge available at the time of Galileo, and they likely would have dropped the notion of the four humors if they had been able to look through van Leeuwenhoek's microscope or had been given one of the first bottles of the Bayer company's new drug.

So Haigh's fundamental error is thinking that the Fathers taught geocentrism as though it were a doctrinal truth, as distinguished from their accepting geocentrism as they accepted other non-doctrinal ideas of their times. That Thomas Aquinas (and other writers in the Middle Ages) also assumed the centrality of the Earth adds nothing to her argument. The praise given Aquinas by popes and scholars has been for his treatises on theology and philosophy, not for his treatises on science—which don't exist. The unanimity of the Fathers, to which Haigh and other geocentrists appeal, simply does not apply here. It applies only to matters of faith and morals, not to matters of science or literature or linguistics or historiography. The attempt to apply the Fathers' common outlook on something non-theological is a misuse of the "unanimous consent of the Fathers."

Back to Galileo. Haigh quotes Andrew Dickson White as saying,

> During nearly seventy years [following Copernicus] the Church authorities evidently thought it best not to stir the matter, and in some cases professors like [Celio] Calganini [1479-1541] were allowed to present the new view purely as a hypothesis. There were, indeed, mutterings from time to time on the theological side, but there was no great demonstration against the system until 1616. Then, when the Copernican doctrine was upheld by Galileo as a truth, and proved to be a truth by his telescope, the book was taken in hand by the Roman curia.

Haigh says, "No one today would defend Dr. White's statement that Galileo 'proved' the truth of the Copernican system 'by his telescope'"—this is quite right; even Galileo did not think so. "Certainly, nothing was proved to Cardinal Bellarmine's satisfaction." This also is true, but it equally is true that Bellarmine seems to have been open to the possibility that future investigations might result in sufficient proof. This is something Haigh seems reluctant to admit. She passes instead to what she considers to be a crucial part of her argument. "But what is most interesting here in Dr. White's present argument is what he says about the *Index*."

She refers to the *Index Librorum Prohibitorum*, or the *Index of Prohibited Books*. White says he has "eleven different editions of the *Index* in my own possession. . . . Nearly all of these declare on their title pages that they are issued by order of the pontiff of the period, and each is prefaced by a special papal bull or letter. See especially the *Index* of 1664, issued under order by Alexander VII."

Later he says, "To make all complete, there was prefixed to the *Index* of the Church, forbidding 'all writings which affirm the motion of the Earth,' a bull signed by the reigning pope which, by virtue of his infallibility as a divinely guided teacher in matters of faith and morals, clinched this condemnation into the consciences of the whole Christian world. . . . The bull confirmed and approved in expressed terms, finally, decisively, and infallibly, the condemnation of 'all books teaching the movement of the Earth and the stability of the Sun.'"

Haigh delights in seeing these words from a non-Catholic. This is just what she claims, but both she and White are wrong. The *Index* and its prefatory documents, whether in the form of bulls or letters from popes, are disciplinary, not definitional. At any one time the *Index* included books on a wide range of subjects. Some of those books later were removed from the *Index*, and some subjects were no longer listed as being off-limits for writers. One such subject precisely was heliocentrism. A pope cannot bind his successors in matters of discipline. One pope can put a title on the *Index*, and another pope can remove it. One pope can prohibit discussion of a certain topic, and a later pope can encourage discussion. Infallibility does not come into play in any of this.

Less than a century after Alexander VII wrote the bull that prefaced the 1664 edition of the *Index*, Benedict XIV lifted restrictions on some books that taught the Earth's motion (Galileo's work remained on the *Index*), and in 1822 Pius VII ratified an 1820 decision by the Holy Office to drop the prohibition on books about that topic. These two popes, having equal authority with Alexander VII, unraveled an earlier pope's disciplinary decision—something popes had been doing for centuries.

Paula Haigh does not see this as a legitimate exercise of papal power but as a craven collapse before the forces of the modern world: "Benedict XIV and Pius VII yielded to the pressures of a rebellious science which had become arrogantly autonomous. The weakness of these popes *vis-a-vis* the firmness of their predecessors says absolutely nothing about the truth or falsity of the science they thus implicitly endorsed. But it says volumes about those areas in which popes may err and still not call into question the promise of our Lord whose vicar they are." A few pages further along she writes, "Subsequent popes have indeed permitted, under pressure of an arrogant and autonomous science, a contrary view to prevail and have thus allowed the authority of Holy Scripture to be undermined." She offers no citations to cases of "pressure" being applied against these popes by scientists, whether arrogant or not.

If a pope can err in lifting a disciplinary decision, then one should argue that it equally is possible for a pope to err in affirming the decision in the first place. The very fact that two popes undid what an earlier pope had done should tell us that there was no exercise of papal infallibility at any stage in the process. If a pope makes formal use of his extraordinary charism, as Pius XII did in 1950 when defining the dogma of the Assumption of Mary, no later pope would try to undo that definition with his own infallible dogmatic definition—because the Holy Spirit would prevent him from doing so. Once an infallible pronouncement is made, the issue is closed, but if later popes reverse a disciplinary decision by an earlier pope, that is a strong indication that the action of that earlier pope did not involve anything beyond his ordinary powers to issue disciplinary decrees. Haigh fails even to address this argument.

She returns to the idea of angels governing the movements of heavenly bodies. Aquinas explains that "the angels are part of the order of the universe; we are situated between the angels and the animals in the hierarchy of being and have more in common with the angels above us than with the animals below us by reason of our intelligence and free will. That the angels should govern the movements of the heavens in a way analogous to the appointed governance of man is eminently reasonable."

Yes, we have more in common—or at least more

important things in common—with angels than with animals (intelligence and free will being more important than
bodily matter), and, yes again, one can make an analogy
between our superintendence of the things of Earth with
the angels' supposed superintendence of worlds other than
ours. To use medieval parlance, we can say it would be "fitting" for angels to guide heavenly bodies, but fittingness
does not necessarily translate into actuality, and the angels'
movement of heavenly bodies becomes less fitting, less necessary if one may put it that way, when a sufficient natural
explanation can be offered. The more adequate the natural
or scientific explanation, the less necessary even to consider
the fittingness alternative. At some point in the transition
plumping for the angelic explanation becomes not so much
a matter of theology or logic but of sentimentality.

Almost in passing, Haigh buttresses her argument
regarding angels with comments about their opposites.
"That the devils live in the center of the Earth is traditional
teaching of the Church." She cites the article on hell from
the 1910 edition of *The Catholic Encyclopedia*. The portion
she quotes states that "no cogent reason has been advanced
for accepting a metaphorical interpretation in preference to
the most natural meaning of the words of Scripture. Hence
theologians generally accept the opinion that hell is really
within the Earth."

There are three things wrong with this quotation.
First, it finds no analogue in later editions of *The Catholic
Encyclopedia* (which no doubt is why Haigh does not cite
them). Second, the writer goes no further than to say that
the generality of theologians of his time thought hell to be
within the Earth, but the Magisterium does not consist of

theologians—a point repeatedly made in recent years by orthodox Catholics who complain about some theologians overstepping their bounds. Third, and perhaps most telling, is that Haigh breaks off the encyclopedia's quotation too early. This is how it continues: "The Church has decided nothing on this subject; hence we may say hell is a definite place; but where it is, we do not know."

In the final third of her essay Haigh returns to the question of the authority of the bull issued by Alexander VII and prefixed to the *Index*. She begins with the story of William W. Roberts, an English priest who opposed having Vatican I issue a definition of papal infallibility. He wanted to dissuade the bishops who were gathered in Rome in 1870 from committing the Church to what he believed was an erroneous belief. Of course Haigh thinks he was wrong to work against having the council define papal infallibility— something it ended up doing—but she gets mileage from the style of Roberts' argument.

In his book *The Pontifical Decrees Against the Doctrine of the Earth's Movement*, he argues that several popes, and Alexander VII in particular, issued infallible decrees on the topic—or, at least, what were intended to be infallible decrees. In later years science proved that the popes were wrong. They had decreed that the Earth was motionless, but science had demonstrated conclusively its motion. Thus the decrees by those popes were not in fact exercises in papal infallibility, even though at the time the popes thought so. The popes did everything they could to formalize their decrees, yet the decrees still fell short—precisely because there is no such charism as papal infallibility. That was the thrust of his argument.

While rejecting Roberts' belief in the non-existence of papal infallibility, Haigh delights in his argument that those seventeenth-century popes believed they were acting infallibly. This comports with her view: "The evidence for papal infallibility rests, then, upon the bull of Alexander VII in 1664. We might refer to this papal bull as an exercise of the extraordinary Magisterium, whereas the decrees of the *Index* and the *Decree of the Inquisition* condemning Galileo were acts of the ordinary Magisterium." She recognizes that individual acts of the ordinary Magisterium, such as disciplinary decrees, are not, in and of themselves, infallible, but papal bulls might be. (Most papal bulls have not claimed to be infallible decisions about anything.) She continues her argument:

> The modern theologians have never addressed the problem posed by this bull of Alexander VII. If they had, they would need to admit its direct authority and search for some subsequent document by a subsequent pope that formally and specifically abrogated (i.e., nullified) the 1664 bull. But no such document has ever been found or produced. The case seems to me exactly parallel with that of the bull *Quo Primum* of Pope St. Pius V by which he established in 1570 the Mass of the Roman Rite as celebrated by the Council of Trent in perpetuity.

This is an interesting—and telling—parallel. In 1969 Pope Paul VI issued the apostolic constitution *Missale Romanum*. It established the vernacular form of the Mass now commonly called the Ordinary Form. Not a few people were disappointed by the pope's action and kept to the old Mass. A small minority of them went so far as to claim that Paul VI exceeded his authority when he tried to put the old rite on the shelf because the matter was out of his hands. The issue had been decided, said these people, in a definitive, binding, and infallible way when Pius V issued *Quo Primum*. After that, no later pope could undo Pius's decision. No substantial changes could be made in the Mass. Or so the argument went, but it was a poor argument because *Quo Primum*, regardless of its forceful language, was a disciplinary decision, and no pope can bind his successors on disciplinary matters.

Haigh sees a parallel: *Quo Primum* was definitive and irreformable, and so the bull of Alexander VII must have been definitive and irreformable. Her argument collapses if one acknowledges (as nearly everyone even within the Traditionalist wing of Catholicism acknowledges) that *Quo Primum* could be superseded by a later pope. If it could be superseded, so could Alexander's bull.

The argument of Haigh's essay comes to a close this way: "In conclusion, we may confidently assert that the Church's judgment of Galileo's heresy was given with the same kind of infallible authority that we recognize in the universal ordinary Magisterium. Disciplinary deviations from the Church's consistent position are one thing; doctrinal deviations are something else. The Galileo case involved the doctrine of the inspiration and inerrancy of Scripture." But in that Haigh is wrong.

To some, the inerrancy of Scripture may have seemed to have been at stake, but it was not, as even Bellarmine implicitly acknowledged when he noted that his problem with Galileo was that the astronomer had *not yet* provided convincing scientific proof of his theory. Bellarmine tellingly did not insist that such proof never could be found—something he would have said if he thought that the Bible's inerrancy was imperiled.

THE STORY THUS FAR 1

I N THE 1980s and 1990s, at least among Catholics, the chief promoters of geocentrism were women. Neither Solange Hertz nor Paula Haigh has claimed knowledge of physics, mathematics, or astronomy. Their approach to cosmology is not scientific but affective. It also might be called disaffective because their embrace of geocentrism is a consequence of their rejection of other things.

In Haigh's mind, biological evolution and heliocentrism are yoked together, and they are to be rejected together. Hertz sees in heliocentrism the culmination of political and social errors that stretch back centuries. She takes an oppositionist attitude toward modernity as a whole, and heliocentrism is modern. Both women see in today's science a Jewish influence that they believe is corrosive of the Catholic faith. This becomes a commonplace among later geocentrists, some of whom veer into outright anti-Semitism. Both women pit earlier Church documents against later ones. In their minds, the old documents hold trump.

What becomes clear, even in these relatively early writings, is that geocentrism—in the minds of those who promote it publicly but also in the minds of everyday adherents—is part of a package. It is not held in isolation. It goes hand in hand with opposition to widely-accepted ideas in biology, geology, and physics, and it is allied to a persistently literalistic view of the Bible.

PART 2

"PRETENDING TO BE ALL SCIENCE-Y"

THE FIRST ANNUAL Catholic Conference on Geocentrism was held in South Bend, Indiana, on November 6, 2010, and consisted of ten lectures punctuated by breaks for lunch, dinner, and a question-and-answer period. Five of the nine speakers were Catholics, and four were Protestants.

Robert Sungenis spoke on "Geocentrism: They Know It But They're Hiding It." His acolyte Mark Wyatt gave an "Introduction to the Mechanics of Geocentrism." (For tax-reporting purposes, Wyatt—who lives in Chino, California, and works for Avery Dennison—is listed as president of Sungenis's Catholic Apologetics International, but in public Sungenis assigns that title to himself.)

Robert Bennett, co-author with Sungenis of *Galileo Was Wrong: The Church Was Right*, to which he contributed about ten percent of the text, spoke on "Scientific Experiments Showing Earth Motionless in Space." Rick DeLano—co-producer with Sungenis of the film *The Principle*—talked about "Scientific Evidence: Earth in the Center of the Universe."

Martin Selbrede spoke on "Answering Common Objections to Geocentrism." He is vice president of the Chalcedon Foundation, a Protestant group that promotes Christian Reconstruction, a position developed by the late Rousas John Rushdoony, who taught that Old Testament laws should be reinstituted by today's civil authorities. Selbrede was followed by the most prolific writer among Protestant geocentrists, Gerardus Bouw, who discussed "The Biblical Firmament: Outer Space is Not Empty."

Sungenis returned to talk about "Galileo and the Church: What Really Happened?" Two more Catholics followed. Best known for writing against Masonry, John Salza talked on "The Fathers and Exegesis of Scripture on Geocentrism." E. Michael Jones, editor of *Culture Wars*, gave the talk that was the most peripheral to the main subject: "English Ideology, Newton, and the Exploitation of Science." Rounding out the event was another Protestant, Hugh Miller, who looked at "Carbon 14 and Radiometric Dating Showing a Young Earth."

About one hundred people attended the conference, which was held at the Hilton Garden Inn, only a six-minute drive from Jones's home. One of those in attendance was Todd Charles Wood, a Protestant working for the Center for Origins Research, a creationist organization affiliated with Bryan College, which was named after William Jennings Bryan and is located in Dayton, Tennessee, site of the 1925 Scopes Trial. Wood studied at Liberty University and the University of Virginia. He has a Ph.D. in biochemistry and serves as president of the Creation Biology Society. Like all the members of that group, he holds to the young-Earth thesis.

At the geocentrism conference Wood met attendees who had come from as far away as Mexico, El Salvador, and Puerto Rico. "Judging by the way some of them responded to the talks, there were a fair number of geocentric enthusiasts in the audience," he says in an online report.[19] "Others seemed to be taking a wait-and-see attitude, open to the idea of geocentrism but not convinced. Others seemed outright skeptical." Skeptical himself, before the conference Wood had worked up "simple objections" to geocentrism, such as "Foucault's pendulum, stellar parallax, red shift, cosmic background radiation, [and] retrograde planetary motion." He was surprised to see that the speakers "already thought of answers. I didn't always find the answers convincing (or even comprehensible), but at least they'd thought about it."

For Wood, "the central issue . . . has always been what the Bible actually teaches about the motion of the Earth (or lack thereof)." That theme was addressed by John Salza, who "kept saying, 'The Bible never says that the Earth is in orbit!' That's not really the point. The Bible doesn't give us the periodic table of the elements either. . . . Given [Salza's] unsatisfying treatment of the passages I'm familiar with, I have my doubts he adequately interpreted the ones I'm not familiar with. In the Q&A, a priest asked whether the Church Fathers affirmed geocentricity because that's what Scripture taught or because that was just the most popular view of the day. I thought that was a really good question,

19. Too Charles Wood, "Hanging Out with the Geocentrists" (Nov. 2010), toddcwood.blogspot.com/2010/11/hanging-out-with-geocentrists-part-1.html.

but Salza didn't answer it." The speakers "didn't convince me that the Bible required this geocentric position."

Wood was not much impressed by Robert Bennett's talk. He had expected better, because he saw that the program listed Bennett as holding a "doctorate in physics with emphasis on Einstein's relativity." Wood thought "this was finally our opportunity to get down to some of the really hard questions about geocentrism." It turned out not to be.

Bennett "opened with about five to ten minutes of discussion of the scoffers and mockers. He welcomed them, but he did come across as kind of miffed at being the object of scorn. Then again, who can blame him? Then he talked about the aether, which is apparently the key to understanding geocentrism." In this Wood is correct. If aether does not exist (and in fact it does not), geocentrism collapses because there is no way to explain how the star field can circle the Earth every twenty-four hours, and there is no way to explain what keeps the Earth at the center of the universe.

Bennett claimed that aether "supposedly explains the motion of Foucault's pendulum, and then he talked about the speed of light" and concluded "by explaining how there is some kind of signal in the cosmic background radiation just behind the constellation Leo." That constellation represents Jesus Christ since Leo is the "Lion of Judah," said Bennett. He claimed that Regulus, a star in Leo, is "therefore signaling to us that Jesus is returning," says Wood. "I'm truly at a loss for words."

Robert Sungenis gave one of the historical talks. It was on Galileo. "Knowing his bias, I didn't expect much from his talk. . . . I was actually surprised by how blatantly anti-Galileo it was. Not just in the way he presented, but in the way Sungenis sort of brushed by the facts of the situation. For example, he described the 'Letter to the Grand Duchess Christina' as 'Galileo goes on and on about why heliocentrism does not contradict the Scripture.' That's not exactly correct. Galileo was proposing a hermeneutic in the 'Letter,' a way of interpreting Scripture that made it possible to be a heliocentrist. His argument was far bigger than just 'why heliocentrism does not contradict the Scripture.'"

Sungenis got other things wrong about the Galileo story. Galileo had "let the manuscript [for his *Dialogue Concerning the Two Chief World Systems*] sit for nine years," but Sungenis "made it sound like Galileo just marched right out defiantly against the Church and began writing his next argument for heliocentrism, but that's not what happened at all."

For most of his lecture Sungenis talked about "what happened after Galileo, in which he emphasized that it was the infallible Magisterium of the Church that condemned Galileo and therefore the condemnation could not be rescinded without admitting that the Magisterium had made a mistake. I found this section interesting, but, given Sungenis's mistreatment of Galileo, I have to say I'm dubious about some of his claims here."

Wood had high hopes for Gerardus Bouw, who holds a doctorate in astronomy. Prior to his talk, Bouw responded to a challenge during the Q&A session: if the universe revolves around Earth daily, Saturn must be moving at the speed of light. "Yes," says Wood. "I had thought of that too." More distant objects would need to travel even faster than Saturn. "Bouw said something like this: superluminal velocity is not a problem for an omnipotent aether plenum." That was it. Sungenis then moved to another questioner.

Later, during his talk, Bouw returned to the issue and "asked us to imagine the inverse of nothing." He said, in Wood's words, that "the inverse of nothing is infinite. Therefore God is the inverse of nothing. And by the way, that doctrine of creation *ex nihilo*? Not biblical. The Catholic crowd had a hard time swallowing that one." At the end of Bouw's talk Sungenis said, "Dr. Bouw has a doctorate in astronomy, so if you think he knows what he's talking about, that's probably why." Wood says, "That's not at all what I was thinking."

In the concluding section of his report on the conference Wood remarks of the speakers: "I personally suspect that they're completely, utterly, and unnecessarily wrong, but I know they're absolutely convinced they're right." He advises the organizers, if they plan another conference, not to cram so many talks into one long day, and he says it is important to "avoid *ad hominem* arguments. For a group that is routinely mocked and held up to public scorn as examples of medieval foolishness, you sure do like to indulge in personal criticisms of your critics. Whatever happened to the Golden Rule? . . . Characterizing scientists as arrogant, deceptive, or purely driven by philosophical bias doesn't help your case at all. It makes you sound like conspiracy kooks. . . . Insulting those who disagree with you doesn't add anything to your argument, and it makes you look bitter and petty."

Wood asks the organizers to "please stop with the hand-waving. Frequently during the conference, speakers would make highly conjectural claims that could have been verified [through calculations or experiments] but weren't. . . . Instead what we got were speculations. Hand-waving. *If* you calculate this, it supports geocentrism. *If* you do that experiment, it supports geocentrism. Enough with the *ifs*! If it's not a hard experiment or calculation, then do it! . . . Stop pretending to be all science-y when you're not."

"I'm done with geocentrism," ends Wood. He says he "wanted to at least consider the possibility that I was wrong about the universe, but at this point I remain totally unconvinced that I am. Perhaps, if the geocentrists take my advice seriously, they might convince a few more people. Or they might just convince themselves that the Earth really does revolve around the Sun."

AN EVANGELICAL WARNS EVANGELICALS

ONE PROTESTANT SCHOLAR and scientist who has sparred with geocentrists is Robert C. Newman, emeritus professor at Biblical Theological Seminary, located near Philadelphia. The school identifies itself as "Evangelical Protestant" and was founded largely for the training of missionaries. In its statement of principles it affirms the Apostles' and Nicene Creeds. It is not a Fundamentalist institution, but it is theologically conservative.

Newman received his Ph.D. in astrophysics from Cornell University. He has written at least two essays against geocentrism. "Geocentrism: Was Galileo Wrong?"[20] carries no date of publication, but the most recent books in its bibliography are from 1993, and other internal evidence indicates the essay could not have been written long after that year. Newman's second essay is titled "Evangelicals

20. Robert C. Newman, "Geocentrism: Was Galileo Wrong?" (Interdisciplinary Bible Research Institute, 1993), ibri.org/Tracts/geocntct. htm.

and Crackpot Science"[21] and is dated March 15, 2000. Let's look at them in turn.

Newman first came across a tract advocating geocentrism in 1967. The author was Walter van der Kamp, and the title was *The Heart of the Matter*. Later a colleague heard of a conference promoting geocentrism and managed to become a speaker at it even though his paper discussed the theory's scientific problems. "He was only able to get his paper accepted for the conference by giving it an ambiguous title and abstract. The title was 'Geocentricity: What Saith the Scriptures?'" Newman continues: "It is hard to say how prevalent this view [geocentrism] is." A then-recent bibliography of 1,850 works against evolution "lists five recent works advocating geocentrism." That was more than a quarter of a century ago. Today, those five works would be joined by ten or twenty more.

Although geocentrism was not then a topic of interest to most Christians—this remains true today—it was not unlikely for the average Christian to be asked about it. Such a person could profit from a few pointers, which Newman hoped to provide in "Geocentrism: Was Galileo Wrong?" His strongest arguments come in a section called "Does the Earth Move Around the Sun?" He seeks to address particularly those geocentrists who agree that the Earth is stationary in space but who say it rotates once daily; this is contrary to the majority opinion among geocentrists. That opinion holds that the Earth not only is stationary but also does not rotate. A consequence of the majority opinion

21. Robert C. Newman, "Evangelicals and Crackpot Science" (Leadership University, 2000), leaderu.com/science/crackpot.html.

is that the rest of the universe must orbit the Earth every twenty-four hours. The minority geocentric opinion necessitates only that the Earth is the center of the universe.

Newman says that, for geocentrists who hold to a rotating Earth, the "easiest-to-understand problem" concerns meteors. He says that, among amateur stargazers, it is "well known . . . that one sees far more meteors after midnight than before midnight":

The reason for this is that (viewed looking down from above the North Pole) the Earth is moving counterclockwise around the Sun and turning counterclockwise on its axis, so that from noon to midnight we are on the back side of the Earth as it travels around the Sun, and from midnight to noon we are on the front side. On the front side, the Earth is hit by meteors coming toward it and may even overtake a few. On the back side, the meteors will not hit the Earth unless they are traveling fast enough to catch up. . . . For the same reason, we have lots of bugs splatter on the windshields of our cars but very few on the rear windows!

Newman turns briefly to Einstein's theory of relativity, saying that geocentrists sometimes argue that the theory indicates that "any reference frame is as good as any other, since all are equivalent. Therefore, their Earth-centered reference frame is just as good as a Sun-centered one, and science cannot distinguish between them. Unfortunately, otherwise competent scientists have sometimes said the same thing. But this is not true." Newman says that "Einstein's principle of relativity refers to frames of reference which are moving at a *constant velocity* in a straight line relative to one another. This is not the case with the Sun and Earth, where the relative motion is approximately circular." (This is the extent of Newman's remarks on relativity.)

Just as police officers can determine motorists' speeds through radar, so speeds of relatively close objects such as asteroids and planets can be determined with radar, but the stars are too far away. Nevertheless, says Newman, "we can tell something about their motion relative to us by two means":

> For closer stars we can detect their sideways motion against the background of more distant starts. For stars close or far, we can detect their motion towards or away from us by a shift in the frequency of the light they emit, the so-called Doppler shift. We find in each of these motions two components, one due to the motion of the Earth around the Sun (which varies with a year-long period) and another due to the relative motion of the Sun and the particular star (which has no such periodicity). Again, clear evidence that the Earth is moving around the Sun.

> These motions of other stars in our galaxy relative to our Sun are consistent with the idea that our Sun and these stars are all rotating around the center of our Milky Way galaxy. Doppler shift measurements of other galaxies such as the Magellanic Clouds or

the Andromeda Galaxy show that their stars are also rotating around the centers of these galaxies.

Newman concludes his short essay by saying, "These evidences indicate that there is no validity to the idea that the Earth is the physical center of the universe. We as Christians have a good reason to believe that the Earth has a central place in God's redemptive plan, but it does not follow from this that our *location* is central. God warned the nation of Israel not to think that because he had chosen them they were particularly great in themselves. The situation here is similar."

This is a good point. The ancient nation of Israel was central to God's plan, but Israel was not centrally located in terms of geography. In ancient times it never was more than a crossroads, a place one might pass through en route to greater cultures, such as those of Egypt or Assyria or Rome. Likewise, there is no reason for a Christian to believe that God's plan, now operating through the New Covenant, requires geographic centrality. If the Earth is central to God's plan (and it is), it is because of *what* the Earth is, not *where* it is.

Newman's second essay, "Evangelicals and Crackpot Science," covers multiple issues, not just geocentrism. His purpose is not to critique each issue at length—the essay is only 6,000 words long—but to provide evidence, through sample bibliographies, that some Evangelicals have been going off onto strange tangents. Among topics not directly related to geocentrism are the notion of a flat Earth, the belief that before the Flood there was an ice canopy (rather than a vapor canopy) surrounding the Earth, the idea that quantum theory is false, various uses of numerology, and the writings of Immanuel Velikovsky (1895-1979), whose books attracted notoriety during the third quarter of the twentieth century.

In Newman's short section on geocentrism he lists nine works promoting the theory, all but one of them having been published within the prior twenty years and all, apparently, by Protestants. He acknowledges that geocentrists who "have the Earth absolutely still would seem to be on more solid ground than the rotating-Earth variety":

> There are several biblical references to the Sun rising, the world not moving, plus Joshua's command for the Sun to stop, but no passages that tell us whether during the course of the year it is the Earth that goes around the Sun or vice versa. I don't have the space here to deal with the hermeneutical assumptions of geocentrists, except to note that they reject the very reasonable suggestion that all these passages are looking at matters from the perspective of one standing on the surface of the Earth rather than of one looking down from space. God does condescend to speak to us in human language.

Some geocentrists argue for a small universe—the position taken by evangelist Harold Camping, for whom Robert Sungenis used to work before returning to the Catholic faith. Usually the small-universe position goes hand-in-hand with believing in a young Earth, though the reverse is not usually true. If everything we see has been created within the last ten thousand years, say those who believe in a small universe, there would be no way for light from distant stars to have reached us if the universe were immensely wide, yet we see the light.

This problem can be overcome by positing a universe that is only a few light-years across. Camping, for example, holds that stars that appear more distant are just smaller and dimmer than other stars, thus giving a false impression of great distance, but Newman notes that it is not possible for stars so small to exist because "their gravity wouldn't be sufficient to hold the hot gases together nor provide high enough temperatures to run their nuclear furnaces."

So what are small-universe people to do? One solution is to argue that the speed of light has not been constant, a view adopted also by some geocentrists who don't think the universe is small. Anyone thinking the cosmos was created thousands of years ago rather than billions of years ago must grapple with the question of how starlight could have reached the Earth in an astronomical twinkling of the eye. Perhaps, ages ago, light moved much faster than it does today, and perhaps modern notions of light's fixed speed are wrong. Newman responds this way:

> But to be able to see objects ten billion light-years away if the universe is only ten thousand years old means that the speed of light since creation must average a million times larger than it is now! If the speed of light were only a thousand times faster in early human history [as asserted by Barry Setterfield of Australia's Creation Science

Association], we should observe some drastic consequences.

Einstein's famous equation $E = mc^2$ relates the conversion of matter to energy, where c is the speed of light. If we keep the masses of objects constant but let c be larger by a thousand back in patriarchal times, then the heat output of the sun and of radioactive elements would be a million times [one thousand squared] what it is now, frying everything in sight!

If instead we require that E be constant to avoid this problem, then the masses of objects will be a million times smaller, and neither humans or air would have been heavy enough to keep from floating away from Earth, and life would have been impossible. Thus we have clear historical evidence that the speed of light has never changed by anything close to what Setterfield needs.

Next Newman has a subsection on the "Astronomical Confirmation of Joshua's Long Day." He looks chiefly at the claim that modern science has been able to verify that there is a "missing day" on the calendar and says that claims "that the lengthened day has been confirmed by astronomical observations, however, appear to be hoaxes. [Harold] Hill's story—reprinted in newspapers throughout the U.S. in the 1970s—is that computers at NASA's Goddard Space Flight Center in Greenbelt, Maryland, detected a missing day in past time and that 23 hours and 20 minutes of it were found at Joshua's time and the other 40 minutes when the Sun went backward ten degrees in Isaiah's day. NASA denies any such discovery." (One might wonder whether Harold Hill's story might have been induced, in part, by his sharing the name of the flimflam man in Meredith Willson's *The Music Man*.)

In actuality, there is no way to determine whether there is a "missing day." To do so requires two sets of data that can be compared, such as precise historical dating and an astronomical extrapolation. Newman gives as an example an eclipse. If historical records give an eclipse's "exact date, time, and place . . . before the time of Joshua," and if, "by calculating back from the present we found that the eclipse 'should' have taken place exactly one day earlier than the historical report says, we would have such a day missing." But the most ancient record we have of an eclipse occurring in the right area dates from 1217 B.C., long after Joshua's time. "There appears to be no way for us to detect the sort of discrepancy Hill alleges"—or for NASA to detect it.

THUMBING ONE'S NOSE

D EATH TO EINSTEIN![22] is the provocative title of a thin book by Scott Reeves. He describes himself as "the author of numerous science fiction and fantasy novels." When he studied at Arizona State University, he majored in biology, and he says he "was born in the same town where Captain Janeway will eventually be born." The reference is to Kathryn Janeway in the *Star Trek: Voyager* television series, which ran from 1995–2001. Her supposed future birthplace is Bloomington, Indiana.

Interestingly, the Janeway character was played by Kate Mulgrew, who served as narrator for Robert Sungenis's 2014 film *The Principle*. When she learned that the film was intended to promote geocentrism—not something evident to her from the lines she was given to read—and that Sungenis, who had a track record as a conspiracist, was the principal behind *The Principle*, she denounced it and him.

Reeves says on his website that his "great ambition is to prove that Albert Einstein was wrong and [that] his theory of relativity has stifled scientific progress for a hundred years."

22. Scott Reeves, *Death to Einstein!* (Aether Wind, 2013).

That goal is repeated in the opening sentence of *Death to Einstein!*: "This is not a book to explain Einstein's theory of relativity to newcomers. This is a book to debunk the theory."

Reeves explains that in the nineteenth century "most scientists believed that light waves traveled through a medium that filled all of space, called the *luminiferous aether*. Much the way sound waves require a medium such as air to propagate, so it was believed that light required a similar medium":

> The Michelson-Morley experiment was intended to detect the motion of the Earth relative to the luminiferous aether. The reasoning behind the experiment was simple. If, as [James Clerk] Maxwell said, light travels at a constant speed through the electromagnetic medium, then, if you're moving relative to the medium, you should be able to detect a change in the speed of light, specifically, the velocity of light you measure should be the speed of light minus your own velocity. The technical details of the experiment aren't important. What is important is that the experiment *failed to detect any motion of the Earth relative to the luminiferous aether* [his emphasis].

This posed a problem for scientists. "Either they were wrong about the Earth moving through space, or there was something peculiar going on that desperately needed to be explained." Reeves says that "at the end of the nineteenth century, there was incontrovertible evidence supporting the view that the Earth was motionless in space. . . . . Scientists completely ignored this 'preposterous' view for decades, searching for an alternate explanation for Michelson-Morley. . . . [T]hey puzzled over the results for decades before Einstein came along and developed relativity, rescuing the scientists from an Earth-centered universe." He adds, "We have two sets of data that unequivocally support an Earth-centered universe: Michelson-Morley, which supports that Earth is not moving, and astronomical observations which support the interpretation that Earth occupies a special place in the universe."

Reeves is so sure this is true that he asserts, "Anyone who wants to believe that Earth is in an absolutely privileged place at the center of the universe is perfectly safe in thumbing his nose at the outraged relativist and crowing, 'Na nah na nah na nah! Can't touch me!' Cries of 'Crackpot!' and 'Lunatic!' are completely unfounded and unjustified and are made from a place of ignorance."

Like other geocentrists, Reeves argues that going from heliocentrism to geocentrism is mathematically easy. Working up the equations that describe celestial motion should be no more difficult for one system than for the other. "Yes, and the motion of the universe around the Earth could be equally well calculated if enough time were devoted to it. . . . The only choice between an Earth-centered universe and a non-Earth-centered universe is a choice between coordinate systems!"

Robert Sungenis, in an article called "The Geocentrism Challenge,"[23] offers "$1,000 to the first person who can

23. Robert Sungenis, "The Geocentrism Challenge" (May 7, 2002), originally appeared at the Catholic Apologetics International website but has been removed.

prove that the Earth revolves around the Sun." He says, "Mathematically speaking, as several astronomers have told me, one could make Jupiter the center of the universe and work out a mathematical model in which all of the motions of the heavenly bodies are accounted for." He gives no evidence that anyone ever has attempted to do this, nor does he name the "several astronomers" he spoke with. Like Reeves, Sungenis seems to think that working out the calculations, while tedious, is doable, but he does not mentioned anyone who has tried to do it.

This raises the question why no geocentrist has taken the time to work out the equations for his system. If geocentrism indeed was proved by the Michelson-Morley experiment, surely someone in the succeeding century-and-a-quarter would have done the geocentric math, at least to show how (in the modified Tychonian system that most geocentrists endorse) the planets revolve around the Sun and the Sun revolves around the Earth. But no one has.

Reeves elsewhere insists that

> The only possible reason scientists might not want the geocentric model presented in textbooks, rather than the heliocentric, is that they don't want people to realize that it is, in fact, just as valid to say that the Sun orbits the Earth as that the Earth orbits the Sun. They don't want such a model presented because it's a slippery slope that leads to claiming that the geocentric frame is the absolute frame. And God, yes God, forbid, we don't want the public sliding down that slope, back into the Dark Ages.

There is a simpler way to explain this reluctance to have "the geocentric model presented in textbooks": its proponents have failed to offer the mathematical calculations that they say will prove it. Until they do, they hardly can object to others—including educators—not accepting their system. The new geocentrists make broad claims, but they fail to show through calculation or experimentation that their claims are justified, and they ignore commonsense alternatives to their theories. For example, all along there has been an easy way to explain the unexpected result of the Michelson-Morley experiment. That solution is to realize that aether simply does not exist.

THE STORY THUS FAR 2

THERE HAS BEEN only one Annual Catholic Conference on Geocentrism, and nearly half of its speakers were non-Catholics. Most proponents of geocentrism, whether Catholic or Protestant, are creationists, so it is noteworthy that the most sustained negative review of the conference came from a creationist. Todd Charles Wood, a Protestant young-Earther, found the speakers "pretending to be all science-y." They employed scientific language without showing competence in science. This will prove to be a recurring pattern: much use of scientific terminology by people who seem incapable of doing the science.

Wood's observation is shared by Robert C. Newman, an astrophysicist who formerly taught at an Evangelical seminary. In a short article he discusses missteps made by new geocentrists not just in science but in exegesis. He debunks the oft-made claim that science has determined that there is a "missing day" that can be traced back to the time of the Bible's Joshua. Newman notes that he first came across a pro-geocentrism tract in 1967, which was a decade-and-a-half before Solange Hertz's first pro-geocentrism article. That tract was by a Protestant geocentrist.

Among geocentrists, Albert Einstein is a *bête noire*. They say he conjured up his two theories of relativity so scientists could escape the real conclusion of the Michelson-Morley experiment, which, in their minds, showed Earth to be immobile and aether to exist. Many geocentrists write against Einstein, but one geocentrist, Scott Reeves, says it is his "great ambition is to prove that Albert Einstein was wrong and [that] his theory of relativity has stifled scientific progress for a hundred years." If the general and special theories of relativity can be thrown out, heliocentrism will be jettisoned too, say many geocentrists, but they are wrong. After all, heliocentrism was the accepted theory long before Einstein's arrival.

PART 3

THE PROTESTANT CHAMPION

GERARDUS BOUW PUBLISHED *A Geocentricity Primer* in 1999 and a revised version in 2004.[24] The book is an abridgment of the original edition of *Geocentricity*,[25] which first appeared in 1994. In its current edition the smaller book is really two books in one: Bouw's 158-page book is paired with Gordon Bane's even shorter *The Geocentric Bible*.[26] There is not much to Bane's book, which is almost exclusively a tendentious reading of biblical passages, so it may be good to consider it first.

Bane is a Protestant who adheres to the King James Only position, which holds that the KJV—also known as the Authorized Version—is the only fully accurate translation into English and that the KJV has a quasi-inspired status—a belief not known to be held by any Christian whose

24. Gerardus Bouw, *A Geocentricity Primer* (Cleveland: The Biblical Astronomer, 2004).
25. Gerardus Bouw, *Geocentricity* (Cleveland: The Biblical Astronomer, 2013 [1994]).
26. Gordon Bane, *The Geocentric Bible* (Cleveland, The Biblical Astronomer, 2004).

first language is other than English (with the exception of Bouw, whose first language was Dutch).

Bane begins with a disclaimer. He notes that Bouw disagrees with some of his ideas, in particular "the small universe model," which holds that the stars are very much closer—and therefore very much smaller—than they generally are thought to be. Positing a universe with a small radius partly resolves a problem facing geocentrism: how to accommodate a star field that circles the Earth daily without the moving stars violating the limit on the speed of light. As will be shown, Bouw, who does not hold to the small-universe model, has what he considers to be an adequate solution.

Like most geocentrists, who face formidable problems of demonstration for their theory, Bane asserts that, when considering heliocentrism and geocentrism, "there is no proof for or against either theory. All tests, including the stationary satellite, can be explained just as easily from the geocentric viewpoint." (This is not really the case. The geocentric argument regarding geostationary satellites falls apart on anything more than a cursory examination.)

Like Bouw, Bane believes in aether, invisible stuff by and through which light propagates. He says aether "can neither be seen, tasted, smelled, touched, nor weighed" and that it "is the substance of which all space beyond the atmosphere consists." In other words, aether can't be detected by any of the senses or by any instrumental extension of the senses, such as electron microscopes or particle accelerators. There is no way to show, experimentally, that aether exists. It is a mental construct and nothing more.

Bane imagines aether to be made of almost infinitesimally

small particles, many orders of magnitude smaller than the smallest subatomic particle so far discovered. "Each particle of aether moves slightly out of position causing a wave, just as each drop of water does when a wave travels across the sea. The existence of aether must be true above the atmosphere, because waves must be waves of *something*." If this does not seem particularly clear, it is because Bane doesn't have much to say about aether. Bouw has more to say, as will be seen, but the result is not more clarity.

In *A Geocentricity Primer* Bouw does not deal with the problem (for geocentrists) of geostationary satellites. Bane doesn't hold back. He writes:

> The most difficult phenomenon to reconcile with a geocentric universe is the stationary satellite at 22,300 [actually, 22,236] miles above the Earth. According to the heliocentric theory, they [sic] travel at the same speed as the Earth rotates, and centrifugal force holds them up. If neither the satellite nor the Earth rotate[s] according to the geocentric theory, what holds them up? It is perfectly possible, and not an illogical concept, that the rotation of the distant masses [of stars] can generate [a] gravitational field that exactly equals the centrifugal field.

This is a variant of the standard geocentric argument, which holds that the Earth's tug on a satellite is balanced by the gravitational pull of distant stars. More consideration of this notion will be given later. Here it is enough to note that if it is the gravitational force of the stars that holds a motionless satellite in place (whether that force derives from the stars' mass or from their rotation), there are equally many stars on the other side of the Earth, adding their pull to its tremendous gravitational force. The stars on the one side would seem to cancel out the stars on the other, leaving only the Earth's gravity to work on the satellite and leaving geocentrists with a difficult-to-explain manifestation of levitation.

After Bane's brief considerations of aether and geostationary satellites, he turns to scriptural exegesis. Almost every page of the remainder of his book consists of scriptural citations and their explication. Occasionally he takes to task Protestant translations other than the King James Version. He says, with respect to problems of cosmology, that any "final solution must be going back to the absolute truth in the King James Bible, which is openly geocentric." This is a common posture for geocentrists, whether Protestant or Catholic. They differ in their preferred translations of the Bible, but nearly all of them assert that science is unable to determine whether geocentrism or heliocentrism is true. Surety on the issue can be obtained only from the sacred text.

Now let's turn to Gerardus Bouw. His undergraduate degree was in astrophysics and his master's degree and Ph.D. were in astronomy. Born in 1945, he is professor emeritus of computer science at Baldwin-Wallace University in Ohio. Writing of himself in the third person, he says that his "intense interest in cosmology led him to a spiritual birth in 1975." When he was young, Bouw believed in theistic evolution. In college he concluded that such a theory was unworkable. He gave up religion "because I recognized that evolution and the Bible don't mix." He was "trying to 'correct' the Bible to fit evolution, and not the

THE PROTESTANT CHAMPION

other way around." About a decade later he "became a special creationist (meaning that the universe is no more than 6,000 years old)." He "joined the Creation Research Society and soon ran into some differences with them because many, though not all the members of that learned society, are scientists first and biblicists second."

Today Bouw is "an ardent believer in the inerrancy and preservation of the Holy Bible." Like Bane, he adheres to the King James Only school of thought. When he was about thirty, Bouw "started a critical reading of the Bible, from cover to cover, searching for inconsistencies and contradictions between an infinite God and the God of the Bible. Fortunately, God was watching out for me in that the only Bible I owned was an Authorized Version, the only English Bible free of such contradictions. Any other version and I would have been left with no alternative but agnosticism."

Bouw cites James Hanson (born 1933), professor of computer science at Cleveland State University, as a formative influence on him in those years. Like Bouw, Hanson is a geocentrist. Bouw says, when they first met, their "main disagreement lay in the inerrancy of the King James Bible. Jim had the quaint notion that it is the inerrant, preserved word of God in English, if not in the world. Jim left me with one final question: 'If the King James Bible is not the word of God, then where is it?' I had no answer to that one, and after a few minutes of thought I was forced to the conclusion that the King James Bible is truly the inerrant, preserved word of God."

Once Bouw was converted on that issue, he and Hanson "immediately organized the first 'geocentricity' conference, which drew speakers from British Columbia to Bulgaria, and from the local schools to the Massachusetts Institute of Technology." This was in 1978. The conference lasted five days and was held at Cleveland State University. "Half-hour papers were presented morning and afternoon. . . . Also present, and extremely outraged that the notion of geocentricity should be so well received, was a graduate student [not identified by Bouw] from Kent State University, who from that day on undertook a personal vendetta against Professor Hanson."

Today Bouw heads the Association for Biblical Astronomy. Its website, geocentricity.com, describes the organization as "a ministry of the Mantua Country Baptist Church," which styles itself "independent" and "fundamental." The small church is located in Aurora, Ohio. The Association for Biblical Astronomy had been founded by Walter van der Kamp in 1971 as the Tychonian Society. Its quarterly journal originally was called *The Bulletin of the Tychonian Society.* Bouw began writing for the publication in 1976. The next year he wrote his first book on geocentrism, *With Every Wind of Doctrine.*[27] "The 500 copies sold out by offering it in the *Bulletin*, but mostly by word of mouth." In 1984 Bouw became editor of the *Bulletin*, and, after van der Kamp's death in 1988, he became head of the organization, which he renamed. The journal also was renamed, to *The Biblical Astronomer.*

Turning back to his interest in Scripture, in 1991 Bouw wrote *The Book of Bible Problems*, which he calls "the first treatise treating the major so-called contradictions of the

27. Gerardus Bouw, *With Every Wind of Doctrine* (Tychonian Society, 1984 [1977]).

Bible." This is incorrect—it was not the first such book. Preceding it were many works with similar titles, such as R. A. Torrey's *Difficulties in the Bible* (1907), William Arndt's *Bible Difficulties* (1932), and Gleason Archer's well-known *Encyclopedia of Bible Difficulties* (1982). Perhaps Bouw means that his was the first such book from a strict KJV-Only perspective, but his phrasing suggests that he did not investigate whether Protestants before him had written similar books. (The genre has been a popular one, particularly in Evangelical circles.)

Bouw's most accessible book on geocentrism is *A Geocentricity Primer*. In its first paragraph he biases the discussion: "Four hundred years ago there raged a debate among the learned men of Europe about whether or not the Earth orbits the Sun." On one side were "biblicists," "who held that the Sun goes around the Earth once a day," and on the other side were "secularists," who "maintained that the Earth daily rotates on an axis." Notice the anachronistic "secularists." In the seventeenth century there were no such people; there was no such term. The term was first used in the middle of the nineteenth century. Bouw has characterized his opposition with an epithet. It comes as no surprise to see that he nowhere states his opponents' best case.

On the opening page of the first chapter Bouw claims that "it took almost a hundred years for heliocentrism to become the dominant opinion, and it did so without any scientific evidence in its favor." None at all?

Geocentrists commonly allege that evidence in favor of heliocentrism equally can be used to prove geocentrism, if only the equations are inverted. Evidence for one cancels out evidence for the other. This isn't true, but this claim is proof that they believe scientific evidence exists—for both sides. Based on the science alone, they say, either side could be right. To determine which in fact is right, we must turn to Scripture. Still, in making these assertions most geocentrists acknowledge that heliocentrists can offer scientific evidence that supports their position, but Bouw does not number himself among such relatively broad-minded geocentrists.

Like other proponents of his position, Bouw sees a slippery slope: "The Copernican Revolution, as this change of view is called, was not just a revolution in astronomy, but it also spread into politics and theology. In particular it set the stage for the development of Bible criticism. After all, if God cannot be taken literally when he writes of the 'rising of the Sun,' then how can he be taken literally in writing of the 'rising of the Son'?" This is a slogan, not an argument, and, like all slogans, it is more false than true. It is little more than a play on words and, if translated into most other languages, falls flat, since the words for Sun and Son are not near homonyms in most languages.

A few pages later Bouw explains a term he claims as his own invention, geocentricity, which he contrasts with geocentrism. In early centuries, "the geocentric model was one where the planets moved on crystalline spheres and where no astral body could leave its particular sphere. Geocentricity, by contrast, is an integrative model which ties the parts of the cosmos together as a whole." It accounts for the more recently observed movements of planets, stars, and sundry other bodies, such as asteroids, that do not remain in any fixed spherical shell. Bouw's new term does not seem particularly useful, and it is not used widely by other proponents of an Earth-centered system. It is noted here but generally will not be used in the remainder of this discussion, except where Bouw is quoted.

The next major section of Bouw's book—at one hundred pages nearly two-thirds of the whole text—offers the reader almost no science but plenty of scriptural exegesis. Bouw's goal is to demonstrate from the Bible that the Earth is motionless. He examines verses such as Psalm 104:5, which says that God "laid the foundations of the Earth, that it should not be removed for ever" (KJV). He argues that "removed" indicates that the Earth does not rotate, but he acknowledges that others argue that the word means the Earth is steady in its orbit around the Sun. He cites other translations. They too are susceptible to varying interpretations, but Bouw thinks he has sufficient grounds to eliminate all interpretations other than his own.

A passage he examines at length is 2 Kings 20:9–11, which refer to a sign given to Hezekiah through Isaiah: a sundial's shadow moving backward ten degrees, which supposedly implies the Sun moving backward—thus the Sun must orbit the Earth. Bouw gathers similar stories from other cultures, calling his sections "Hezekiah's Sign in India," "Hezekiah's Sign in China," "Hezekiah's Sign in North America," and so on. He assigns rough dates to these stories and concludes that each ancient culture witnessed the same biblical event. He doesn't give much consideration to whether the sign given to Hezekiah could have been a geographically localized occurrence for the king's benefit or whether other cultures may have had memories of entirely unrelated, and probably fabulous, occurrences.

The same sort of argument is built for "Joshua's long day," as recounted in Joshua 10:12–14. Bouw collects similar stories from other cultures and offers "Joshua's Long Day in Africa," "The Chinese Account of Joshua's Long Day," "Joshua's Long Day in North America," and so on. In some of these regions what was a long day for Joshua would have been a long night for the locals, given that they lived thousands of miles from Canaan.

Bouw dates the biblical event to 1448 B.C., but the best he can say regarding the distant cultures is that they have similar stories that cannot be dated: "The Ojibways tell of a long night without any light. The Wyandot Indians told missionary Paul Le Jeune of a long night. The Dogrib Indians of the Northwest tell of a day when the Sun was caught at noon and it instantly became dark." Even though not all of the North America stories involve a long night (to correspond with a long day for Joshua), they are suffi-cient for Bouw to conclude that the "preponderance of long night tales in the Americas would rule out the theory that Joshua's long day was a miracle which was local to Canaan."

Not so. Assuming these disparate cultures really do have legends of a long day or a long night, nothing in what Bouw presents leads one to think that those legends go back 3,500 years or so. There is no way to date any of them. Some may be older than that; others may have arisen only a few generations ago, particularly those that exist only as oral accounts (which is to say probably all of them). Bouw relies on the existence of such tales to rule out a sensible explana-tion for Joshua's long day: that it was a localized phenom-enon, a vision given to those present at a particular place at a particular time and not something universally observed. A good parallel might be the Miracle of the Sun at Fatima.

On October 13, 1917, a crowd estimated to range

between 30,000 and 100,000 gathered in the Cova da Iria, a field near Fatima, Portugal. For several months three young visionaries had been claiming to see the Virgin Mary, who, they said, had told them that a miracle would be performed that day. In those years Portugal labored under an anti-clerical regime; anti-clerical newspaper reporters were on hand to report what they presumed would be a non-event. Those reporters were religious skeptics and had no reason to promote what they considered to be superstition.

That morning had seen substantial rain. The field and the people were soaked. Then, according to witness Almeida Garrett, a professor of natural sciences at Coimbra University,

> The Sun's disk did not remain immobile. This was not the sparkling of a heavenly body, for it spun round on itself in a mad whirl, when suddenly a clamor was heard from all the people. The Sun, whirling, seemed to loosen itself from the firmament and advance threateningly upon the Earth as if to crush us with its huge fiery weight. The sensation during those moments was terrible.[28]

The vision lasted about ten minutes. At its conclusion, the ground and clothes were reported as dry.

I am not suggesting that the Miracle of the Sun at Fatima is on the same level as a miracle reported in the Bible. The Catholic Church does not require her members

28. John De Marchi, *The Immaculate Heart: The True Story of Our Lady of Fatima* (New York: Farrar, Strauss & Young, 1956), 146.

to believe either in the messages given in a private apparition or even in the apparition's occurrence, but the large number of witnesses who claim to have seen essentially the same thing is a strong indication that something inexplicable occurred. (Some have argued that the Miracle of the Sun was an example of mass hysteria, but there is no recorded incident of even ten people going hysteric at once, let alone tens of thousands—including non-religious reporters for anti-clerical newspapers.)

What the people at the Cova da Iria saw, however it might be understood, was a local occurrence. They were convinced that they saw the Sun moving wildly in the sky, but elsewhere in Europe and around the world nothing remarkable was reported.

(Let me note at least one exception to this. Around thirty years ago I met an elderly Portuguese priest. He related that, when he was a child, he had seen the Miracle of the Sun—but not at Fatima. He was living in the Azores, islands that form an autonomous region of Portugal and that lie about 850 miles west of the Portuguese mainland. He told his family what he had seen, but no one believed him. They chalked it up to childish fantasizing. A few weeks later came news of what had occurred at Fatima, and the adults realized that the boy's description fit the news reports.)

To sum up: there is good evidence, even if not conclusive to an anti-miraculist, for the Miracle of the Sun at Fatima. The people present in that field thought they saw the Sun moving erratically through the sky, but the event was local. It did not involve the actual displacement of the Sun, though observers imagined it did. Likewise, the event involving Joshua could have been local; the Sun in that instance also could have had an apparent but not actual displacement. If understood this way, "Joshua's long day" tells us nothing whatsoever about the relative motion of the Earth and the Sun.

After Bouw examines a few other Bible verses, including some that are said to support heliocentrism (he quickly dismisses these), he turns again to matters of science, but first he says the equivalent of what Bane says: "We have also noted that absolute proof of either heliocentrism or geocentricity is lacking unless one accepts the testimony of the Bible as such an absolute proof." For him as for Bane, the answer to the scientific question is one of biblical exegesis, not of mathematics or physics.

After explaining several variants of the geocentric theory, Bouw brings up "the firmament model." He acknowledges that in the nineteenth and early twentieth centuries experiments seemed to do away with the notion of aether, but now "the concept of the aether is coming back into vogue." Those experiments had tried to establish the existence of a "luminiferous aether," an extremely rarefied substratum that allows light waves to propagate, much as ocean waves propagate in water and sound waves in air. In the end, the experiments were unable to show that such an aether exists.

But, says Bouw, there is another kind of aether that he calls the plenum. Instead of being almost infinitely rarefied, it is almost infinitely dense. This aether is made up of Planck particles, which are fantastically smaller than any subatomic particle. Bouw thinks "we are engulfed in a sea of Planck particles." They make up virtually the entire mass of the universe. The plenum is effectively a complete solid.

(In *Geocentricity* Bouw claims that one cubic centimeter of the firmament "is more massive than ten trillion trillion trillion universes."[29]) "One would think, for example, that it would be impossible to move in such a medium, just as one could not move if encased in iron—even if one were made of solid iron! Normally this is true," but the wavelengths of nuclear particles "are so long compared to that of the Planck particles that the plenum is transparent to them."

If that sounds unclear, it is because it is. Bouw does not so much prove the existence of this plenum as he asserts it. Its existence is necessary for him because of a problem that arises in the geocentric model. Geocentrists hold that the entire universe rotates daily around a stationary and non-rotating Earth. The further a body is from the Earth, the faster it must travel to make the circuit in twenty-four hours. A near planet, such as Mars, would have to travel fast, but a distant planet, such as Neptune, would have to travel many times faster because it has a much longer distance to cover in the same amount of time. Saturn is roughly at the distance at which a body would have to travel faster than the speed of light to make a daily circuit. The immensely more distant stars would have to travel countless times the speed of light—but nothing can travel faster than light, right?

No, says Bouw. The key is that light's speed limit is defined relative to the plenum, which Bouw identifies with the firmament mentioned in Genesis. The firmament itself rotates around the Earth: "Because of its tremendous mass and density compared to the material universe, it is a small

29. *Geocentrcity*, 418.

thing for the firmament to rotate once a day." The stars and other bodies are carried along by the firmament and "seldom go much faster relative to the firmament than a few hundreds to a few thousands of miles per second, far, far below the speed of light." Thus, those distant stars, which would have to violate the speed of light many times over if they traveled through absolutely empty space, do not violate it because space is not empty. It is filled with the plenum, which in the speed of its distant parts is not limited by the speed of light since it is far more fundamental than light itself.

The problem for Bouw is that there is no evidence that this plenum aether, which he identifies with the firmament of Genesis, exists. He works backwards from a premise that the universe rotates around the Earth daily, and he concludes that distant bodies must move in some medium that accommodates the limitation on the speed of light. It is special pleading and nothing more.

There are two points that Bouw does not address. He thinks the Earth is motionless, with the entire universe revolving around it each day. He fails to note that this is equivalent to the universe being motionless with the Earth rotating on its axis each day. There is no way to distinguish one situation from the other. This alternate understanding, with the Earth rotating and the universe not, removes any need to consider what happens if a body exceeds the speed of light.

The other point that Bouw does not address is that if, as he says, the plenum drags along the distant stars, it also should drag along planets, comets, and other objects relatively near to us, but in his telling these bodies travel through the plenum without interference. This suggests that the distant plenum has one set of properties while the near plenum has another. How could that be, if the plenum is uniform throughout space? The discrepancy cannot be accounted for by the size of the objects. Bouw could not say that small objects—planet-sized and below—can travel unimpeded through the plenum while massive objects such as stars are pulled along it. If that were so, stars would find themselves stripped of their planets.

Bouw concludes *A Geocentricity Primer* with a consideration of "the moral effects of heliocentrism." In an earlier chapter he had said that

> the [Protestant] Reformers were correct in their moral trepidations about heliocentrism. . . . [T]he concern of the Reformers and other Christians has proven to be well founded; for heliocentrism directly spawned the view that man is but a mere machine, a cosmic accident. Heliocentrism is widely acknowledged [by whom?] as the foundation of the impersonal, mechanistic, materialistic universe and the existentialist view that human life is purposeless and thus, by implication, worthless.

In Bouw's mind, philosophic materialism arose in consequence of the triumph of heliocentrism. He makes no reference to Greek philosophers such as Thales, Anaxagoras, Epicurus, and Democritus, who laid down what later would be termed materialistic principles, nor does he refer to Lucretius, whose *De Rerum Natura* (first century B.C.)

makes use of the view that everything that exists is matter and void.

More sophisticated than Bouw's view of materialism is Arnold Lunn's. In the preface to *The Flight From Reason* he says, "The doctrine that reality is describable in terms of matter and motion, that quality is capable of being expressed in terms of quantity, though an ancient error, deserves to be described as the Victorian heresy, for it was during the Victorian period that materialism ceased to be the fad of the few and was accepted as the working hypothesis of orthodox science."[30]

Lunn wrote in 1931. The quaint views of Victorian materialists have been revived in recent years, though without the piquancy that sometimes accompanied older formulations. No longer can we find so confident a materialistic explanation as this one, written by Dennis Hird (1850-1920), a British clergyman and writer: "Just as the same particles of matter may at one time form parts of a rose, and at another time parts of a mushroom, so the same force may at one time strike a church as lightning and at another time may be the mother-love that rocks the cradle."[31] Who today, even among ardent materialists, could say a mother's love of her child is really an electrical discharge?

Writing a long lifetime ago, Lunn said that "materialism, in the strict sense of the term, is dead or dying [but] that habit of mind which made materialism possible is still alive today. If we can diagnose correctly the mental malady which was responsible for the Victorian heresy, we shall

30. Arnold Lunn, *The Flight from Reason* (New York: Dial Press, 1931), vii.
31. Dennis Hird, *An Easy Outline of Evolution* (London: Watts, 1903), 184.

be in a better position to understand modern sources of infection." Materialism certainly is still with us, though in a form often more sophisticated (but sometimes not) than the Victorian version that Lunn found himself still having to deal with. Today materialism most commonly is found in the comments and writings of physical scientists, many of whom are learned in their professional fields and yet almost entirely at sea in matters of philosophy and theology. It is in their predecessors that Bouw finds the origins of today's problems.

To him, the great culprits are from the sixteenth and early seventeenth centuries. "Copernicus had succeeded in making a clearly heretical teaching palatable to not only the Roman Catholic Church but to Protestantism as well. Copernicus and Galileo had succeeded in discrediting the Bible as an authority in the realm of science. This called into question the authority of the Bible in other areas, too." This undermining of the Bible may be true for Evangelicals and Fundamentalists, but it is not true for Catholics because the Church never has held that the Bible is "an authority in the realm of science."

Bouw overstates his case when he continues: "After the Galileo affair, the Bible was no longer considered authoritative in the realms of science, philosophy, and day-to-day reality. Less than 200 years after surrendering the Bible's authority in the realm of physical science, man surrendered its spiritual authority at the hands of the German school of higher criticism, a way of criticizing the Bible which supposedly is based on natural revelation, that is, upon 'scientific' principles."

Later, "The revolution of the sciences spilled over

into the political realm. Both the American and French Revolutionary wars stemmed more or less directly from the Copernican Revolution." Bouw presents the reader with no substantiation for this ipse dixit. Like Solange Hertz, whom he does not cite, Bouw dislikes the American political system. He connects his dislike to an astronomical system that was proposed about two centuries before that political system came into existence. For him, post hoc, propter hoc suffices.

Bouw asserts that "heliocentrism's removal of the Bible as absolute authority paved the way for the acceptance of the political lies of evolution and Marxism into man's worldview. The result gave man a lower view of himself and forced him to frame for himself ill-structured questions which can have no answers. Such is the legacy of modern heliocentric science." Here Bouw draws *A Geocentricity Primer* to a close.

MULTIPLE MUDDLES

G ERARDUS BOUW IS a Fundamentalist, but he has a private and unflattering definition of that word and doesn't use it for himself. Nevertheless, his arguments show him to be firmly within the Fundamentalist strain of Protestantism. Like nearly all Fundamentalists, he has little sympathy for and, consequently, a weak understanding of Catholicism, both in its doctrinal and historical aspects. In *Geocentricity*, the revised edition of which appeared in 2013, the few references he makes to the Catholic Church and to Catholic thinking suggest that he has made no serious attempt to understand the religion from which his own faith—whether he realizes it or not—is descended.

(As I wrote in *Catholicism and Fundamentalism*,[32] in popular culture—and particularly in the media— "Fundamentalist" is used negatively, but it should not be. It is the proper label for a certain segment of the Protestant spectrum, and it is a label happily used by many people within that spectrum to describe themselves. They believe

32. Karl Keating, *Catholicism and Fundamentalism* (San Francisco: Ignatius Press, 1988).

that they adhere strictly to the fundamentals of the Christian faith and that other Christians, even Evangelicals, have accepted corruptions of that faith. Whenever I use the term "Fundamentalist," I use it neutrally, not as a pejorative.)

In his discussion of the Galileo case Bouw makes a detour and discusses Vatican I's decree on papal infallibility, a doctrine he of course rejects. He explains the teaching this way: "the pope can make no error when he speaks *ex cathedra* (from the throne) in his capacity as the *Pastor Aeternus* (Eternal Pastor)." A footnote leads the reader to this reference: "Anonymous, 1966. 'Bij de Sluiting van de Concilie,' *Tot Vrijheid Geroepen*, 12(2):29, Feb." The article and journal title translate as "At the Close of the Council" and *Called to Freedom*, a reference to Galatians 5:13: "For, brethren, ye have been called unto liberty; only use not liberty for an occasion to the flesh, but by love serve one another" (KJV). The journal is written from a Reformed— that is, Calvinist—perspective.

The footnote gives insufficient information to judge whether it is the anonymous writer or Bouw himself who misunderstands who the *Pastor Aeternus* is. Likely it is both. At the least the error indicates that Bouw made no attempt to read the dogmatic constitution. Had he done so, he could not have asserted that the *Pastor Aeternus* is the pope. The very first sentence of the document shows that the *Pastor Aeternus* is Christ himself: "The Eternal Pastor and bishop of our souls, in order to continue for all time the life-giving work of his redemption, determined to build up the Holy Church, in which, as the house of the living God, all who believe might be united in the bond of one faith and one charity." The phrase "the life-giving work of his redemption" should have been a clue to Bouw, who happily affirms that there is only one Redeemer—and that he is not the pope. (Calvinists and Catholics agree on this point.)

Particularly egregious misunderstandings on Bouw's part are found in his chapter on "The Reformation and Heliocentrism." He begins by saying the Reformation not only "was a limited return to biblical authority" but "was also a humanistic movement." He asserts that

> humanism* has been a strong faction in the Roman Catholic Church since at least the early thirteenth century, if not since its inception under Constantine. Technically, a humanist is anyone who disavows the deity of our Lord Jesus Christ, maintaining only the Lord's humanity. Today the term, humanism, has lost that clear definition, but in the Middle Ages the original concept of humanism was till intact.

This is a muddle. The Catholic Church did not begin with Constantine. That emperor legalized Christianity in the Roman Empire with his Edict of Milan (313), issued a call for the first ecumenical council (Nicaea, 325), and ended his life as a baptized Christian, but the Church did not begin with him, nor was he ever considered to be its head. This confusion is common among Fundamentalists, for most of whom Christian history ends at the end of the first century and resumes only at the beginning of the sixteenth, the intervening centuries being *terra incognita*. A greater concern here than the date of the establishment of the Catholic Church is Bouw's confusion about humanism.

Today the term generally refers to a secular and often anti-religious movement that disavows the supernatural completely. That is not how the term was used in the sixteenth century and in earlier centuries, when a humanist was understood to be someone who, on the one hand, sought a recovery of Classical learning and, on the other, advocated an active secular but nevertheless Christian life—that is, one outside monastery walls.

The asterisk in the quotation from Bouw leads to a footnote that carries no reference. The writing style of the footnote differs from Bouw's, and the footnote's words seem to have been taken, without attribution, from *The American Heritage Dictionary of the English Language*: "By humanism is meant the cultural and intellectual movement of the Renaissance that emphasized secular concerns as a result of the rediscovery and study of the literature, art, philosophies, sciences, and civilization of ancient Greece and Rome."

This definition contrasts markedly with the usage in Bouw's main text, where he says that "a humanist is anyone who disavows the deity of our Lord Jesus Christ, maintaining only the Lord's humanity." He says that nowadays this clear definition has been abandoned, "but in the Middle Ages the original concept of humanism was still intact."

Indeed it was, but his definition wasn't what the word meant. Christian humanists of the Middle Ages and the Renaissance—Erasmus comes to mind as a good example—held to Catholic teaching. They didn't "disavow the deity of our Lord Jesus Christ." The people who maintained that Christ had only a human nature but not a divine nature were those who subscribed to Arianism, a heresy that flourished in the fourth century, largely died out by the seventh century (though elements persisted through the Middle Ages), but resurfaced in the sixteenth century in consequence of the Reformation.

Bouw—misunderstanding who humanists were—claims there

> were two groups of humanists in Roman Catholicism during the Middle Ages: one group worshiped the Queen of Heaven while the other wanted to eliminate all remembrance of church and deities together. Just before the Reformation began, the two factions were highly polarized, even though Aquinas' works were an attempt at placating them. But the ideas of Aquinas could only carry humanism so far and no farther.

> As soon as the Reformation happened, and as soon as it became evident that the Roman Catholic Church was too weak to assassinate Luther, the dissident humanists seized their opportunity and declared their independence from "mother church" by latching onto the coattails of the Reformers.

This is another muddle. Who is denominated the Queen of Heaven? The Virgin Mary, of course. Who, during the Middle Ages, worshiped her? No one, actually, though Bouw, like many of his co-religionists in the

Fundamentalist wing of Protestantism, believes otherwise. Fundamentalists hold that any honor given to a saint amounts to worship. They are unable to distinguish veneration from adoration, largely because their understanding of the terms is unclear. They are in a position similar to that of a child who is incapable of distinguishing liking from loving. To the child the one is simply a weaker form of the other; in his mind the two are of the same species. To the Fundamentalist, veneration is a weak form of adoration but is adoration nonetheless. This is a limitation in his theology and in his imagination.

Bouw is wrong in thinking that the humanists of the Middle Ages and Renaissance can be divided into two groups, one that worshiped Mary and one that rejected God altogether. Neither group, as such, existed at all—certainly not worshipers of Mary, and there were few people in the Middle Ages "who wanted to eliminate all remembrance of church and deities together." Such people were regarded not as humanists but as outright heretics. In any case, Thomas Aquinas did not write in opposition to these alleged groups, and he certainly didn't try to "placate" them. This assertion is so odd that one must conclude that Bouw never has read anything written by the Angelic Doctor. No one who glances at the table of contents of the *Summa Theologiae* or of the *Summa Contra Gentiles* would write such a thing.

What about Bouw's implication that the leadership of the Catholic Church wished to assassinate Martin Luther but was too weak to do so? Assassinations were hardly unknown in those centuries, and it often did not take much to effect one, even inadvertently. Consider how Henry II's outburst ("Will no one rid me of this turbulent priest?") resulted in the martyrdom of Thomas á Becket in 1170. Does Bouw really believe that the Church lacked the ability to assassinate Luther, had it wished to do so? His accusation likely arises less from his studies than from anti-Catholic sentiments he has imbibed at his church. He seems incurious when it comes to the history and teachings of the Catholic Church. His insouciance with respect to things Catholic may suggest carelessness in his writings about things cosmological.

Bouw summarizes his understanding of humanists in the preface to *Geocentricity*, where he says

The lie that the humanists devised to promote their god over the God of the Holy Bible was subtle enough to deceive all the world's religions, including today's Fundamentalist scholars, since its inception. The lie started to work early in the sixteenth century when a Polish cleric, Nicolaus Copernicus, discovered that a third-century B.C. Greek philosopher named Aristarchus of Samos proposed that the Earth was not at rest at the center of the universe but that, instead, the Sun was located at the center of the universe. . . .

Copernicus, however, saw in Aristarchus' supposition the downfall of Scripture, and, knowing that his idea was heretical to Christianity, sought an occasion to publish it without endangering or inconveniencing his life with the Church of Rome.

Copernicus did so when he knew his death was imminent in 1543.

It was all downhill from there. Today "the churches and their members still profess faith in the word of God, the Holy Bible. But these days such a profession means absolutely nothing. There are more than 230 different Bible versions in circulation in the United States; not one of them professes to be Scripture given by inspiration of God." (This probably will be news to the publishers of those editions.) "The Bible was kicked out of almost every facet of life. It was kicked out of our economy in 1913 resulting in the establishment of the Federal Reserve. America's entertainment industry never had any respect for Scripture." (So much for Cecil B. DeMille's *The Ten Commandments*.) "Our government rid itself of the Bible in 1962, and our schools told God to 'get lost' in 1963." (Actually, it was the Supreme Court that did that.)

That was not the half of it, because "before God was eliminated from those listed institutions, God's word, the Holy Bible, was expelled from this nation's churches. That happened in 1901, when with much fanfare and acceptance, even among the laity, the American Standard Version was introduced to replace the so-called 'obsolete, most inaccurate, archaic' Authorized Version commonly called 'the King James Version.'"

Bouw acknowledges that this is the true subject of his book. "It documents how humanists . . . tricked the churches, their clergy, and their members into abandoning the authority of Scripture in the natural realm. This was accomplished by convincing Bible-believing Christians that some doctrines are essential and others are not. Once that was accepted by the believers, they were ready to abandon what was claimed to be a minor doctrine of Scripture," geocentrism. He ends his preface by saying, "This book is written for you who are truly born-again believers. If you love sound Bible doctrine, this book is for you; it will edify you and strengthen your faith."

In his first chapter Bouw lists his assumptions. Not a single one is about science. All concern Scripture. This is the most important: "I list the assumptions I labor under when it comes to handling the Holy Bible. I assume that Scripture was written by and preserved by an omnipotent, omniscient, and omnipresent God."[33]

In the following chapter he uses similar words: "The reverberations of the Copernican Revolution still ring today, particularly in the realms of politics and theology; for without said revolution, there could be no higher criticism which assumes God is incapable of writing what he meant to say or meaning what he wrote."[34] He repeats this idea at the end of his eleventh chapter: "Again, the issue boils down to the same point we have noted in previous chapters. Either God meant what he wrote or he did not mean what he wrote and would, presumably, revise his original wording if he were to write the passage today. And if he would recant today of what he wrote in times past, then where is truth?"[35]

Again, in the penultimate paragraph of the eighteenth chapter: "Either God means what he has written, or he does not. If God does not mean what he writes or writes what

33. *Geocentricity*, 6.
34. Ibid., 9.
35. Ibid., 19.

he means, then how can he be taken seriously? If God does not inspire literal truth when he mentions the *rising* of the sun, then how can he be taken seriously when he writes of the *rising* of the Son? Without the doctrine of geocentricity, the Gospel is wide open to the charge that it is nothing more than an allegory or fable."[36]

Bouw's second-most important assumption, as given in his first chapter, is that the "standard Scripture"—that is, Scripture that perfectly preserves the inspiration of the autographs—"must always be detectable at all times. Today, the standard Scripture, as proven by the fact that virtually every new version feels obligated to compare itself to it, is the Authorized Version."[37]

Another key point is this: "Without the Copernican Revolution there would be no Marxism in which the state replaces God. Nor could there be any evolutionism with its bigotry and racism and faith that man will eventually evolve to ultimately overpower God."[38] He takes a slightly softer tone later in the book:

Now I do not claim that heliocentrism is primarily responsible for man's moral dilemma today, but its acceptance did pave the way for a worldview which denigrated absolute moral authority to be subservient to man's limited, fallible mind. Heliocentrism's removal of the Bible as absolute authority paved the way for the acceptance of the political lies of evolution and Marxism into man's worldview.

36. Ibid., 242.
37. Ibid., 7.
38. Ibid., 9.

The quotations in the last four paragraphs deserve commentary, even if not extensive. They should not go unchallenged.

Those collected in the first paragraph show that Bouw has a highly anthropomorphic view of how Scripture was composed. He repeatedly says that God "wrote" the sacred text, apparently in the most literal way, but the only place in Scripture where this legitimately might be claimed is in the making of the stone tablets at Mt. Sinai. All the remaining words of Scripture were set down by men. The Catholic Church always has taught that the sacred writers wrote under inspiration, but they were not automatons, nor did they take dictation, except perhaps in a few passages that refer to visions.

For the most part, the writers seemed not to be aware, while they were writing, that they were under the superintendence of the Holy Spirit in terms of producing their texts. Not a single book of the New Testament, other than Revelation, contains anything that suggests that the writer understood himself to be an amanuensis of the divine. The same can be said of most of the books of the Old Testament. In 2 Timothy 4:13, Paul asks the recipient of his letter to do him a favor: "The cloak that I left at Troas with Carpus, when thou comest, bring with thee, and the books, but especially the parchments" (KJV). This, certainly, is something Paul himself composed, not something the Holy Spirit took pen in hand to write down. Bouw fails to appreciate sufficiently that God works through human instruments, even when those human instruments are unconscious of his so working.

As mentioned earlier, and as shown in the second paragraph, Bouw subscribes to the King James Only school of thought. This is very much a minority position even within Fundamentalism. The proponents of the idea never go much beyond claiming its truth, and it is an attractive truth, if true: we have, in our own language, the very words of God, preserved for us as they have been preserved for no others.

It takes but a moment to see that no one outside the English-speaking world would think the King James Only notion makes any sense. Why should God have chosen to guarantee only this translation and not, say, Luther's German translation or translations into French, Spanish, or Italian? The Authorized Version legitimately has been esteemed as a formative influence on the English of the last four centuries, but, like every other translation, it has limitations, both in terms of accuracy of meaning and of felicity of rendering, though followers of the King James Only theory would deny that.

When canvassing the woes of the present era, particularly intellectual and moral woes, one naturally is drawn to trace their origin. In great part that origin will be found in original sin and concupiscence, which is to say at the very origin of the human race. But what about later factors, over which our not-so-remote ancestors had a considerable degree of control? Where did the wrong turn take place? Those against whom Augustine wrote *The City of God* said that the decline of civilization could be traced to Christianity itself. Augustine took care of that argument handily, if not briefly. More recent writers have traced the wrong turn to later times. The American rhetorician Richard Weaver thought the key event was the triumph, in the philosophy of the Middle Ages, of nominalism over realism. Not a few Catholic writers have argued that the great mistake occurred at the Reformation, when Christendom lost its intellectual and spiritual unity.

Geocentrists such as Bouw maintain that the turning point was the acceptance of heliocentrism. They are convinced that the Bible teaches geocentrism as a truth to be accepted as avidly as a Christian accepts any element of the creed. Thinking that God must lie if he permits a sacred writer to write in terms of appearances rather than in terms of brute fact—for the writer to write in terms of the Sun's rising when what is really meant is the Earth's setting—they put a straitjacket on God. They seem oblivious to the many scriptural passages, such as the Song of Songs and much of the wisdom literature, that include verses that no one takes in a crassly literal way but that *could* be taken in such a way, if a reader felt that by doing so he would protect the text.

From the geocentrists' literary error flows their historical error: the notion that from heliocentrism have come our time's chief ideological errors, whether actual or perceived. Geocentrists invest heliocentrism with far more influence than any astronomical theory possibly could have.

Bouw's thirty-ninth chapter is "Geocentrists and Their Critics." One might think this is where his opponents' best cases are given space, but not so. One of the main arguments against a motionless Earth is the geostationary satellite. The heliocentrist thinks he can explain why a satellite can appear to be parked over one spot: it is in orbit around the Earth, but its orbital speed matches the Earth's rotation. The satellite never gets ahead of, and never falls behind, the Earth. It constantly is on the move, but so is the Earth, and so the satellite seems to be stationary—which it is, relative to the ground but not relative to the space through which it is hurtling.

This is not a trivial argument against geocentrism, yet Bouw devotes only a page to it, and that page consists not of his own words but of words from that famous writer Anonymous (who persists in using "geosynchronous" for "geostationary"):

> The difficulty of placing a body in "geosynchronous orbit" is merely that of finding the area of relative gravitational equilibration between Earth and the other bodies of the universe. Since *synchronous* is a symmetrical, transitive, and reflexive relation, a "geosynchronous" body is synchronous with all and only "geosynchronous" bodies. And since the other stellar bodies, of which a "geosynchronous" body is also a satellite are not themselves "geosynchronous," the area of relative gravitational equilibration wanders away from the position occupied by a "geosynchronous" body. Being no longer gravitationally equilibrated, the body loses its "geosynchronicity," and the non-geocentrists says, "Aha! Orbital decay!"[39]

This paragraph is a mess, but it can be translated into clearer English. In Bouw's mind, and in the mind of those geocentrists who say the Earth doesn't rotate on its axis (some geocentrists say it does rotate), what keeps a satellite fixed above a certain spot on the Earth is the pull of gravity from other celestial bodies. Bouw doesn't identify which bodies he has in mind, but he and most other geocentrists

39. Ibid., 658.

think the countervailing pull comes from the stars that are "behind" the satellite.

The obvious problem with this solution is that there equally are stars "in front of" the satellite—that is, on the same side as and "behind" the Earth. The gravitational pull from the stars on that side would cancel out the gravitational pull from the stars "behind" the satellite, leaving it suspended in space with no apparent means of support. Bouw doesn't even bother to address this argument— perhaps the strongest argument against the notion of a motionless Earth.

MEETING TODAY'S GEOCENTRISTS

I N THE MIDDLE of *Geocentricity* Gerardus Bouw devotes a chapter to "Geocentrists from 1650 to 1950." He gives brief biographies of two dozen people, mostly German, American, and English. Closer to the end of the book he gives even more space to an equal number of "Modern Geocentrists," himself included.

Most of those profiled in this later chapter are Protestants, chiefly Fundamentalists and creationists (the two categories do not coincide precisely). Perhaps surprisingly, three Jews are listed. One is Amnon Goldberg (born 1957), whose geocentric beliefs are based on his reading of the Kabbalah. Goldberg has posted thousands of comments on Internet forums and apparently has been banned from some of them repeatedly. A particularly unexpected Jewish name is that of Menachem Schneerson (1902-1994), leader of the Lubavitcher movement. At his death some of his followers expected him to be revealed as the Messiah, a revelation that has not yet occurred. Bouw reproduces an exchange of letters from 1975 in which Schneerson affirms his own geocentric proclivities. The third Jew listed by Bouw is David Lifschultz (born 1945) whom Bouw describes as "a direct descendant from Aaron" and a Karaite, a member of a small branch of Judaism that rejects the Talmud as authoritative and accepts only the Tanakh or Hebrew Bible. Lifschultz has written for *The Biblical Astronomer* and "regards the Authorized Version as the best translation of the Hebrew text into English."

Among those profiled in this chapter, only one clearly is listed as Catholic: Robert Sungenis (born 1955). Bouw says that "Sungenis was raised in a Catholic family and converted to Protestantism when nineteen years old. As a Protestant, he stayed mainline except for the two years he worked for the cult leader Harold Camping's Family Radio Network." Camping (1921-2013) was best known for serially predicting the End Times. His highly-touted prediction that the Rapture would occur in 1994 found him changing the year to 2011 when 1994 got too close. Seemingly undisturbed by his failed prognostications, Camping gained a worldwide following and large annual revenues ($36 million in 2012, the year before his death).

Sungenis, who returned to the Catholic Church in 1992, wrote a book about Camping called *Shockwave 2000!: The Harold Camping 1994 Debacle*. Camping's version of Protestantism emphasized *gnosis* or secret knowledge, though he did not use that term. He thought he was able to divine from Scripture things that had eluded all prior commentators. Camping saw himself as something of a prophet, even if not in the formal sense of the word. Sungenis has come to adopt a similar attitude about himself regarding geocentrism.

Sungenis and Bouw are curious allies. Bouw opposes

Catholic distinctives, which is not surprising given his firm Fundamentalist convictions. Usually people at his end of the religious spectrum refuse to have anything to do with Catholics, but he has found utility in the adage that "the enemy of my enemy is my friend." For his part, Sungenis, when he was active in regular Catholic apologetics in the 1990s, wrote often against Fundamentalist claims. That did not stop him, years later, from inviting Bouw to be a speaker at the First Annual Catholic Conference on Geocentrism, held in 2010 (and not held again since). In fact, of the nine speakers at that conference, four were Protestant, a display of ecumenical latitudinarianism on Sungenis's part.

Early in *Geocentricity* Bouw says that in the second half of the twentieth century geocentrism saw a resurgence. "Currently, three worldwide organizations serve the geo-centric community. All three are mathematically sophisti-cated and have Ph.D.s on their boards, if not as directors." (Critics of these groups argue that being "mathematically sophisticated" is precisely what they are not.)

Of the three groups, the oldest is Bouw's Association for Biblical Astronomy, which was founded by Walter van der Kamp as the Tychonian Society. Bouw makes a point of saying that the organization "is under the directorship of Gerardus D. Bouw, who has an earned Ph.D. in astronomy." He adds that "the Association's geocentric stance is based entirely on Scripture although it can argue on evidence and scientific grounds too." This is a revealing admission but not a surprising one once *Geocentricity* has been read. The overall argument of the book indeed is based mainly on Scripture, not "on evidence and scientific grounds," although those are not absent.

Then Bouw refers to two Catholic groups. The first is the *Cercle Scientifique et Historique* (CESHE), which has offices in Belgium and France. Bouw's group and CESHE "differ on whether the Earth rotates on its axis and [on] the size of the universe. CESHE believes that the Earth rotates and that the universe is small; the ABA believes that the Earth does not rotate and that the universe can be as large as modern science believes it to be. CESHE is devoutly Roman Catholic and was organized to promote the works of Fr. Fernand Crombette."

According to CESHE's webite,[40] Crombette (1880-1970) was not a priest but an eccentric lay scholar. When helping his daughter with a school project, he concluded, from reading Psalm 74, that Jerusalem literally was the center of the world. This led him to study and expand upon Alfred Wegener's theory of continental drift (now, in its developed form, the theory of plate tectonics). Wegener, an exact contemporary of Crombette, proposed in 1912 that the seven continents once formed one land mass that split apart, each section drifting its own way across the face of the globe until reaching today's configuration. Crombette worked twelve years on determining how the various pieces moved and how the continents once fit together. According to an article at the CESHE website, "The result confounded all expectation. The single continent that emerged had the regular form of a flower of eight petals with Jerusalem at its center."

This primitive continent commonly is called Pangaea, but its form, as generally accepted by researchers in plate tectonics, bears little resemblance to Crombette's

40. Ceshe.fr.

reconstruction. The spot that eventually would be the location of Jerusalem is not at the center of Pangaea, which does not look anything like "a flower of eight petals," but is off at one corner.

Later Crombette turned his interest to Egyptian hieroglyphics. He decided that Jean-François Champollion (1790-1832) was wrong in his translation of the Rosetta Stone. The ancient Egyptians did not speak ancient Egyptian but Coptic. (Crombette also thought "that the original language of the Old Testament was Coptic" rather than Hebrew.) With this as his principle, Crombette wrote the fifteen-volume *Book of the Names of the Kings of Egypt*. Following that, "after a long study of geology, and using ancient onomastics [the meaning of names] and toponymy [the study of place names], he wrote the history of the antediluvian patriarchs and then the history of Noah's sons up to the division of lands after Babel."

Crombette "realized to what degree revelation, translated by means of ancient Coptic, shed light on the observations of the secular sciences and gave a coherent and Christian vision of human history." He suspected that the condemnation of Galileo was justified, and his translation into Coptic of some psalms led him to conclude that the geocentric model was correct. He ended with the idea that Jerusalem not only was the center of the world but the center of the universe.

CESHE is a reflection of its founder's idiosyncratic methodologies and conclusions. It promotes geocentrism but as something secondary to wider purposes. The emphasis at the third geocentric organization listed by Bouw is different. Geocentrism is not a secondary conclusion but, today, nearly the sole focus of Robert Sungenis's Catholic Apologetics International, the name of which Bouw gets wrong. He calls the group *Galileo Was Wrong*, but that is the title of Sungenis's geocentric magnum opus. Bouw says that CAI "is Catholic and founds itself on the teachings of the Abbess Hildegard von Bingen (1098-1179)."

While there is a chapter in *Galileo Was Wrong* that refers to Hildegard's vision of the cosmos, it would be a stretch to say that Sungenis's organization was founded on her teachings. It began as an apologetics group that had nothing to do with geocentrism. Later its emphasis was twinned: half on geocentrism and half on exposing supposed Jewish conspiracies in politics and history. Since 2013 the focus has been almost exclusively on geocentrism.

In one respect Bouw is more correct about Sungenis than Sungenis is about himself. Bouw says that Catholic Apologetics International (still the legal name of the organization, even though Sungenis's former bishop wanted him to drop the word "Catholic") was "founded and directed" by Sungenis. The verb "directed," if taken in the sense of "managed," is accurate. Sungenis manages the organization, which in practice is hardly more than a one-man operation.

According to filings with the IRS, Sungenis is not an officer, a member of the board of directors, or even an employee of CAI. The president is Mark Wyatt, who is one of four directors, the others being John Schmieding, Rick DeLano, and Maureen Sungenis, Robert's wife, who also is listed as secretary and treasurer. None of the directors is compensated by the organization.

In recent years, Robert Sungenis has been listed on IRS Form 990 only as an independent contractor, not as an employee. On the return for 2010, for example, his "type of service" is given as "consulting, writing, managing," with his compensation as $168,560. There is no explanation why he isn't listed as president of the organization or as a compensated employee—even though he otherwise holds himself out as president and makes all the decisions for the organization.

At CAI's website he describes himself as "the founder and president of Catholic Apologetics International Publishing, a non-profit corporation," and in online discussions he refers to himself as president of CAI (and he doesn't correct others who refer to him as such and doesn't point them toward Wyatt). Nevertheless, CAI's tax returns list Wyatt as president and Sungenis—when he is listed at all—as an independent contractor. Sungenis has offered no explanation for this anomaly.

Perhaps not surprisingly, most of the people Bouw mentions in the chapter on "Modern Geocentrists" are Protestants. One of them is closely affiliated with Sungenis: Robert Bennett (born 1940). He has a Ph.D. in physics, the topic of his dissertation being rigid body motion in General Relativity. He came to geocentrism not long after Sungenis did. Bouw reports that Bennett's

> interest in geocentricity started about 2005 while reading a dialogue at the Catholic Apologetics International website. Until then he was unaware that the Bible clearly supported geocentricity, so he researched the scientific arguments against geocentricity and "found them sadly lacking." Bob

Bennett's chief contribution to geocentricity thus far is the chapter he wrote for the first volume of Bob Sungenis's *Galileo Was Wrong*. His chapter was the last, most technical chapter in the book.

Actually, Bennett's adoption of geocentrism came earlier. By October 2004 he already had worked up a pro-geocentrism paper for a conference. His chapter in *Galileo Was Wrong*, "Technical and Summary Analysis of Geocentrism," no longer is the last chapter in the work and now appears in the second volume of the three-volume 2014 edition. That chapter occupies 231 pages. (Bennett also contributed the following chapter, "Absolute Lab Frame & Flexible Aether." It runs twenty-seven pages.)

Bouw "inherited" the Tychonian Society from Walter van der Kamp (1913-1998). Born in Holland, van der Kamp moved to Canada, where he became principal of an independent school for Dutch immigrants. He was a member of a Reformed church. In 1963 he joined the Bible-Science Association and the Creation Research Society. The more he studied the early chapters of Genesis, the more he became convinced that creationists were inconsistent in their treatment of the text. They "stoutly denounced" biological evolution, but they refused to accept what seemed to van der Kamp to be clear teaching on the centrality and immobility of the Earth. In 1968 he published a monograph called *The Heart of the Matter*. It was his first public foray into geocentrism.

Bouw reports that "Walter offered it at $2.00 per copy, but few sales went to other than concerned family and friends. Still, Walter felt that there were logical considerations that would not allow him to give up that easily. Walter was optimistic that logic would prevail." A couple of years later the monograph received favorable notice from another geocentrist, and that was enough to induce van der Kamp to form "a most informal and unincorporated organization called the Tychonian Society. With the Society came the *Bulletin of the Tychonian Society*. The first few issues were handwritten and reproduced on a Gestetner. . . . In 1984 Walter handed over the *Bulletin*'s editorial duties to his friend, Gerry Bouw."

Like many geocentrists, van der Kamp maintained ideas that were out of the geocentric mainstream. One was his insistence on a small universe. He thought this had to be the case if Revelation 6:13 were to be upheld. That verse says that "the stars of heaven fell unto the Earth, even as a fig tree casteth her untimely figs, when she is shaken by a mighty wind." If the stars were very far away, they would have to be very large to appear as they do to our eyes—far too large to fall onto the Earth in any meaningful sense. The stars would need to be smaller than the Earth, and that means they would have to be quite close to us. That induces a host of other problems, such as whether the stars we see even could exist as stars if their diameters were measured in hundreds of miles rather than in hundreds of thousands of miles.

A brief mention may be made of Richard G. Elmendorf (born 1927), an engineer who has been a minor player in geocentrism. Bouw reports that in 1978 Elmendorf "issued a $5,000 challenge to anyone who could provide proof of evolution. In 1980 he challenged heliocentrists with $1,000 (later increased to $10,000) for proof-positive of the Earth's motion. Both challenges remain unclaimed." Bouw fails to explain the ground rules. Who was to determine whether a challenge was met successfully—an independent panel composed of people with no particular stake in the issue or just Elmendorf himself? This is reminiscent of a challenge made in 2002 by Robert Sungenis:

> [Catholic Apologetics International] will write a check for $1,000 to the first person who can prove that the Earth revolves around the Sun. (If you lose, then we ask that you make a donation to the apostolate of CAI). Obviously, we at CAI don't think anyone *can* prove it, and thus we can offer such a generous reward. In fact, we may up the ante in the near future. [It never was upped.]

> You can submit your "proofs" to our e-mail address cairomeo@aol.com. We will then offer a response. Both your "proof" and our response will be posted on the CAI science page at our website. If you do not want your actual name listed, we will change your name, but your contents will be posted. If you do not want either your name or your contents posted, then you are not eligible for a reply from CAI nor the $1,000 reward.

> CAI will be the sole judge of whether you have successfully proven your case. But since CAI is built on its reputation of honesty and truthfulness, rest assured that if you do indeed prove your case,

you will be rewarded the money. Now a word of caution. By "proof" we mean that your explanations must be direct, observable, physical, natural, repeatable, unambiguous, and comprehensive. We don't want hearsay, popular opinion, "expert" testimony, majority vote, personal conviction, organizational rulings, superficial analogies, appeals to "simplicity," "apologies" to Galileo, or any other indirect means of persuasion which do not qualify as scientific proof.[41]

At least two men, Ken Cole and J. Lawrence Case, submitted responses to Sungenis's challenge.[42] It seemed to come as no surprise to them or to anyone else that Sungenis declared their arguments to be unconvincing. What is a bit surprising is why Sungenis failed to increased the size of the prize, as Elmendorf did for his. He could have matched the Protestant in ostensible generosity by raising his offer to $10,000, or he could have promised any amount above that. A more generous prize may have induced more people to participate and thus made for a more interesting contest. Sungenis need not have worried about having to mortgage his home because he, as the sole judge, knew *a priori* that any submitted argument would be declared insufficient

41. Robert Sungenis, "The Geocentrism Challenge" (May 7, 2002), originally appeared at the Catholic Apologetics International website but has been removed. The quoted portion of his challenge has been preserved online at biblelight.net/kepler.htm.
42. Cole's argument is found at biblicalcatholic.com/apologetics/GeocentrismDisproved.htm. Case's argument formerly appeared at Sungenis's website, catholicintl.com, but has been removed.

because his theory of the Earth's immobility was, in his mind, itself immovable.

Bouw next devotes a bit more than a page to Martin G. Selbrede (born 1956), whom he describes as "an autodidact" who interned at CalTech and "spent two years at Harvey Mudd College," a school with a strong science curriculum. He subsequently worked for companies that produced flat-panel displays. All along, Selbrede "never stopped learning physics." Beginning in 1980 he became associated with the Chalcedon Foundation and in 2003 became its vice president. In an issue published in 1982, the organization's journal included an article "that mentioned the *Bulletin of the Tychonian Society*. Skeptical but curious to see how well the geocentric case could be made, Martin subscribed and within three years had become a compelling advocate for the minority view."

In 2010 Selbrede was a speaker at Robert Sungenis's First Annual Catholic Conference on Geocentrism. His topic was "Answering Common Objections to Geocentrism." He might have been considered a curious choice as a speaker at a Catholic conference, given his association with the Chalcedon Foundation. In the January 1995 issue of *First Things*, the late Richard John Neuhaus included this description of the movement behind the foundation:

> From time to time, we have had occasion to comment on the Reconstructionists or, as they are sometimes called, theonomists. . . . The most prominent figure in this movement that advocates the universal applicability of Old Testament law is R. J.

Rushdoony, who publishes *Chalcedon Report* out of California. Reconstructionism, in various forms, is not without influence among activists and thinkers in "the religious right." The movement is famously fractious, and for years Rushdoony has been crossing swords with his prolific son-in-law Gary North over questions wondrously exotic.

The latest battle is over "geocentrism," the doctrine that the Earth is the physical center of the creation. Rushdoony is for it and North is against it. Apparently they are lining up astrophysicists and other scientists to help confirm their conflicting exegesis of Scripture. . . . Martin G. Selbrede has a "Chalcedon Position Paper" in defense of geocentricism in the October 1994 *Chalcedon Report*. . . . If we understand the argument, which is doubtful, Einstein is solidly on the Rushdoony side. Galileo has not been heard from, and reports that Gary North is in contact with him remain unconfirmed.[43]

A little further along in *Geocentricity* Bouw devotes several pages to an English engineer, Malcolm Bowden (born 1933). One day, in 1971, Bowden, then something of a theological liberal, listened to a talk by a leader of an environmental organization. The man claimed the world would run out of oil in thirty years and that chaos inevitably would follow. Nothing could avert this outcome,

43. Firstthings.com/article/1995/01/a-pope-of-the-first-millennium-at-the-threshold-of-the-third.

except possibly extensive recycling. "Soon the meeting closed and Bowden drove back the thirty-minute journey to his home. It was during this drive that Malcolm thought about what he had heard and realized that the fight for dwindling oil resources would herald international chaos and wars. Indeed, he knew enough of the Bible to realize that Armageddon could take place within his lifetime." The drive home also was revelatory to Bowden in an entirely unexpected way.

He insists that it was definitely not a vision but a picture that formed within his mind. In his mind he saw a white-robed man, about a mile tall and to whom Bowden felt about the size of an ant, looking straight ahead as if looking into the distant future. The giant in Bowden's mind spoke some reassuring words, which Malcolm reiterates were not audible except within his mind. He said, "Do not worry; everything is under my control and everything is going according to my plan." Then the mental picture faded and Bowden arrived home.

Looking for an explanation of this event, "Bowden began to read a pocket New Testament to fill the void. He knew instinctively that this would ultimately provide the true pathway to follow. Through his reading, Malcolm came to realize that his liberal Methodist Church was clueless to explain what had happened to him. He knew it involved God in a big way, but that was about all he knew."

Eventually Bowden joined a theologically conservative church and became a creationist.

Bouw reports that Bowden "was contacted by a very intelligent and well-informed Roman Catholic who had written several very interesting papers that he sent to Malcolm." They included "very controversial articles about finance, the one world order, politics, relativity, Einstein, etc. One paper was about geocentricity." He gave little weight to its arguments, but later he "obtained Bouw's books and realized that, just like the theory of evolution, heliocentrism was only supported by mass media propaganda and little else. It was Bouw's books that convinced him that the scientific evidence supported geocentricity."

The convert to geocentrism promoted his new faith "in creationist forums, where he thought the evidence that supported the Bible would be welcomed. To his surprise, he was met with ridicule and anger. Bowden was accused of 'bringing the creationist movement into disrepute.'" In short order he was removed from participation in several creationist online venues. "Because of how hostile creationists are to geocentricity, Bowden has not (as yet) given a full lecture on the subject but would be willing do so if requested by interested people." Bowden thus seems to be among the most minor of minor characters in Bouw's cast of promoters of geocentrism—he gives no public talks, has written no books on the topic (though he has written four on creation), and has not even managed to maintain a pro-geocentric presence online.

Why does Bouw even discuss him? Not, surely, out of sympathy for Bowden's vision of a mile-tall Christ; in Bouw's segment of Christianity, such things are not regarded kindly because any private apparition—whether just mental or otherwise—is thought to undermine the finality of the revelation that ended with the writing of the last book of the New Testament. So why does Bouw mention Bowden, someone inconsequential to the geocentric movement? Perhaps it has to do with Bowden's acknowledgment of the influence on him of Bouw's books.

Of the remaining "modern geocentrists" listed by Bouw, the final one to mention here is Marshall Hall (born 1930). The reader may remember his name from the chapter about Solange Hertz, where she is quoted as praising his book *The Earth Is Not Moving*:[44] "a quintessentially popular treatment of this difficult subject, and he must be given much credit for bringing the arcana of modern mathematical physics down to the level of us scientifically illiterate mortals." Hertz the Catholic has a higher opinion of Hall than does Bouw the Protestant.

Bouw met Hall in 1984, and a few years later Hall began reading the works of geocentrists such as "Walter van der Kamp, Jim Hanson, Dick Elmendorf, and yours truly." Hall said, "These works heightened the suspicions I already had on this subject and led to some years of study and finally resulted in a book in 1991 (*The Earth Is Not Moving*)."

Bouw doesn't think much of the book, in which "Hall focuses on conspiracies; he believes that physicists and mathematicians conspire against the geocentric truth. He documents the satanic influence of Kepler starting from [Kepler's] youth on. Marshall was one of the first to suspect that Kepler may have murdered his mentor, Tycho Brahe."

44. This is in Paula Haigh's "Galileo's Heresy."

Referring to Marshall's website, Hall says, "the purely assumption-based Copernican model has historically provided the keystone of today's science-controlling Big Bang evolutionary paradigm. This, in turn, has led to facts which reveal the Kabbalist sources responsible for every concept which makes up that evolutionary paradigm."

Bouw thinks that "the main drawback with Marshall Hall's conspiratorial approach is that, from his supporting documentation it is at times impossible to tell where one authority leaves off and another begins. . . . When it comes to Scripture, Marshall suffers from conspiracy overload." Bouw finds this tendency off-putting, but it may have been Hall's conspiratorialist mindset that made his writings attractive to Solange Hertz. For Hall the conspiracies were chiefly undertaken by Jews; for Hertz, the culprits were Freemasons.

Two pages after Bouw's profile of Hall comes his shorter look at Robert Sungenis, who, as will be explained later, has subscribed to innumerable conspiracy theories, most of them involving Jews and some of them involving Freemasons. Bouw does not advert to his Catholic friend's proclivities.

A CREATIONIST ATTACKS
GEOCENTRISM

D ANNY FAULKNER IS on the staff of Answers in Genesis, a Protestant group that describes itself as "the world's largest apologetics organization." It may well be, with a budget of about $20 million and several hundred employees. Its focus is creationism, a literal understanding of the early chapters of Genesis, and an opposition to the theory of biological evolution. Faulkner joined the organization in 2013, after retiring from 25 years as professor at the University of South Carolina's Lancaster campus, where he taught physics and astronomy, the subjects of his master's degree and Ph.D., which he obtained from Clemson University and Indiana University.

He says, "I was born again at age six as the result of a vacation Bible school." He fell away but his faith revived when he was a teenager. He came across *The Bible and Modern Science* by Henry M. Morris, the founder of the Institute for Creation Research, now headquartered in Dallas. The book confirmed Faulkner in his young-Earth creationist beliefs, which he now is able to pursue more avidly at Answers in Genesis.

Most creationists do not subscribe to geocentrism. Faulkner seems to wish that none of them did: "Alas, there are recent creationists in the world who are geocentrists. They teach that the rejection of God's Word did not begin with Darwin's theory of biological evolution or even with Hutton and Lyell's geological uniformitarianism. Instead, they argue that the scientific rebellion against God began much earlier with heliocentrism."[45] He notes,

> Perhaps the best-known geocentrist in the world today is Gerardus Bouw. . . . To distinguish modern geocentrism from ancient geocentrism, Bouw has coined the term "geocentricity" for the former. Bouw has a Ph.D. in astronomy from Case Western Reserve University,[46] so he certainly is in a position to know and understand the issues and literature involved. Given Bouw's stature as the chief champion of geocentricity, we will use his book by the same name as the primary source on the topic.

These words come from the opening pages of "Geocentrism and Creation," a 10,000-word article written by Faulkner in 2001. It is divided into four main sections: biblical issues, historical issues, scientific issues, and

45. Danny Faulkner, "Geocentrism and Creation" (Aug. 1, 2001), creation.com/geocentrism-and-creation.
46. Ironically, this is the institution at which Lawrence Krauss, who is interviewed in the pro-geocentrism film *The Principle*, taught from 1993–2005. Bouw studied there beginning in 1973.

an appendix on the geocentric and heliocentric models. At least in one respect things have changed since the article first appeared: while Bouw remains a major figure in promoting geocentrism, he no longer can be considered "the chief champion of geocentricity," no matter which name is used for the theory. That mantle today resides elsewhere, as will be discussed later.

Faulkner takes a less ideological approach to Scripture than does Bouw, saying:

> There are few biblical texts that in any way even remotely address the heliocentric/geocentric question. In each instance there is considerable doubt as to whether cosmology is the issue. Some of these verses are in the poetic books, such as the Psalms. It is poor practice to build any teaching or doctrine solely or primarily upon passages from the poetic books, though they can amplify concepts clearly taught elsewhere. It is also important not to base doctrines upon any passage that at best only remotely addresses an issue. That is, if cosmology is clearly not the point of a passage, then extracting cosmological meaning can be very dangerous.

This is a danger that Gerardus Bouw and, like him, Gordon Bane see not at all. They cobble together every scriptural passage that might be construed in any way, poetic or otherwise, as having a reference to the cosmos. They don't allow poetry to be poetry or everyday expressions, such as "the Sun sets," to be anything other than oracles of science, though even those who fervently hold to heliocentrism speak in terms of the Sun and the Moon rising and setting. Faulkner says,

> Much of the case for geocentrism relies upon many biblical passages that refer to sunrise and sunset. Geocentrists argue that since the Bible is inspired by God, then when he chose to use such terminology, the Lord must mean that the Sun moves. By this reasoning, virtually all astronomers and astronomical books and magazines are geocentric, because "sunrise" and "sunset" is exactly the language that such sources use.

> Anyone who has spent much time watching the sky can testify that each day the Sun, Moon, planets, and most stars do rise, move across the sky, and then set. Such observation and description do not at all address what actually causes this motion. However, the geocentrists will have none of it, insisting that language and usage must conform to their standards.

Faulkner notes that Bouw spends considerable time explicating verses such as Psalm 93:1: "the world also is stablished, that it cannot be moved" (KJV). "He claims that 'stablish' is the proper translation as opposed to 'establish' that is used in modern translations. He states that the former word means to stabilize, while the latter means to set up. However, none of the English dictionaries (including the Oxford) I consulted support this distinction. All of the dictionaries revealed that 'stablish' is an archaic variation of 'establish.'"

Faulkner discusses Bouw's reliance on the Hebrew that lies behind "stablish." Bouw claims the usage is determinative, but Faulkner notes that the same Hebrew word is translated almost everywhere else in the King James Version of the Old Testament as "establish."

More to the point, Faulkner faults Bouw's argument that verses such as Psalm 93:1 indicate that the world does not move. "These passages declare that the world is not to be moved" by "an external or causative agent to bring about change in position," such as moving the Earth out of its orbit. The phrasing "does not exclude the possibility of motion apart from an external agent." In other words, Psalm 93:1 and like passages do not exclude Earth's orbiting the Sun or its diurnal rotation.

In his section about historical issues, Faulkner says that "Bouw claims that heliocentrism has led to all sorts of moral degeneracy." One example Bouw discusses is a supposed link between heliocentrism and astrology. "This is a bizarre assertion," says Faulkner, "given that astrology flourished in the millennia before the heliocentric theory became popular and seems to have *decreased* where heliocentrism has flourished. Ironically, the dominant geocentric theory of history, the Ptolemaic system, was devised primarily as a tool to calculate planetary positions in the past and future as an aid for astrological prognostications."

Faulkner turns to Bouw's criticism of Nicolaus Copernicus's skills. "According to Bouw, Copernicus had the time to spend investigating alternate cosmological models because Copernicus was not gifted enough to be in demand for astrological calculations"—that is, Copernicus had meager mathematical skills. "With Bouw, Copernicus cannot win—if he had done horoscopes, Bouw would have castigated him as a mystic dabbling in the occult, but since he did not do horoscopes, it was because Copernicus was a poor mathematician." Something similar could be said about Tycho Brahe and Johannes Kepler:

> Apparently it has never occurred to Bouw that the reason that Tycho was available to pursue astronomical measurements rather than produce horoscopes may have been the same reason that he claimed that Copernicus had time to pursue other matters. Indeed, late in life, Tycho realized that he was not the best mathematician around and needed help in making sense of his observations. This caused Tycho to seek the best mathematician available, who happened to be Kepler.

Faulkner thinks Bouw works on the basis of a double standard, holding heliocentrism to a standard to which he doesn't hold geocentrism. Bouw argues "against the legitimacy of heliocentricity, because it was prematurely accepted before there was any evidence. Yes, he also admits that by 1650 there was no solid proof for or against either the heliocentric or Tychonian models." If Bouw were consistent, "we should reject both models in favor of the Ptolemaic model or some other alternative, but of course Bouw insists that only the heliocentric model be subjected to such scrutiny."

The discussion turns to Galileo. He discovered four moons of Jupiter and "used this to counter the objection to heliocentrism that the Moon would be left behind if the

Earth moved. It is obvious that Jupiter moves, and it is also obvious that its motion does not leave behind the satellites of Jupiter. Bouw is correct that this is an argument by analogy, but one cannot so easily dismiss this argument. The critics of heliocentrism must explain how the motions of Jupiter and its moons and the Earth and its Moon are different."

In Galileo's time the predominant geocentric theory held that all celestial objects directly orbited the Earth. This Ptolemaic model was undercut when Galileo demonstrated that some objects—the moons of Jupiter—directly orbited some other body and only indirectly orbited the Earth. The Tychonian model of geocentrism, in which only the Moon and the Sun orbit the Earth and all other objects orbit the Sun, except for various planets' satellites, set this objection aside. Geocentrism had to be modified to "save the appearance," and the appearances had changed once the telescope came into use.

Faulkner accuses Bouw of not paying sufficient attention to the timeline. Bouw claims that seventeenth-century heliocentrists were unable to offer proofs of their theory. "In the absence of a real challenge to the status quo," as Faulkner puts it, "the status quo should prevail." Bouw notes that the status quo was geocentrism, and this means that his preferred system, the Tychonian, should prevail, but Faulkner notes that the Tychonian model was not the status quo at that time. The Ptolemaic model was. He says Bouw "slips his model in as a substitute."

Faulkner then turns to two key scientific matters, parallax and star streaming.

Parallax is the apparent displacement of an object when viewed from two different locations. If the Earth orbits the

Sun, one would expect to see a displacement of a particular star if that star were observed when the Earth was on opposite sides of the Sun. An object closer to Earth will have a larger angle of displacement than will a more distant object. The observed angle assists in determining how far away the object is.

This is how human vision works. The eyes are slightly separated from one another (the analogue is the Earth at opposites sides of its orbit). When we view an object, our minds use parallax to get depth perception. A smaller angle is interpreted by our minds to mean the object is more distant. With an object essentially at infinity—say, a star—there is almost no angle formed; the eyes look out almost in parallel. With an object at the tip of the nose, we have to look at it cross-eyed, which makes for an angle approaching 90 degrees.

Faulkner admits that Bouw is correct in noting "that the failure to detect stellar parallax was an argument against the heliocentric model." Like geocentrists of the seventeenth century, Bouw says this is at least an indirect argument in favor of geocentrism. True, but the later discovery of stellar parallax, says Faulkner, equally should be regarded as substantiation of heliocentrism. That later discovery did not occur until 1838; it was accomplished by a German astronomer, Friedrich Bessel.

Why did the verification take two centuries? Because in Galileo's time there were no instruments fine enough to record extremely small angles. Even the nearest stars are so distant than the subtended angle, measured when Earth is at opposites sides of its orbit, is equivalent to the angle formed by the outside edges of an object two centimeters (less than one inch) in diameter that is placed 5.3 kilometers (about three miles) from the viewer. Faulkner says Bouw tries to get around this by modifying his version of the Tychonian model "so that the Sun in its annual motion drags along the distant stars." It is not clear how the Sun, which is not a large star, manages to move stars that are incredibly distant and, many of them, immensely larger than itself.

Then Faulkner turns to star streaming. The term has multiple meanings, but he uses it to refer to the motion of stars relative to the Sun:

> The Sun is moving through space, as can be deduced by proper motions (the gradual motion of the stars across the sky) of many stars. The first measurement of this was done more than two centuries ago by the great German-born English astronomer William Herschel (1738-1822), though the measurement has been refined many times since then. When the proper motions of many stars are considered, we find that stars seem to stream out of a region called the solar apex, presumably in the direction in which the Sun is moving.

> Conversely, stars appear to stream toward a convergent point, called the solar antepex, diametrically opposed from the solar apex and presumed to be the direction from which the Sun is moving. This would appear to be strong evidence that neither the Sun nor the Earth is the center of the universe, but Bouw baldly asserts that stars could be moving past the Sun rather than the other way around.

In his conclusion, Faulkner writes that the adoption of the heliocentric theory was a result of the application of Occam's razor, a principle developed by the medieval philosopher and theologian William of Ockham (or Occam). The principle holds that, among competing hypotheses, the one requiring the fewest assumptions should be preferred, all else being equal. "The Sun-centered system was far simpler than the primary geocentric model, the Ptolemaic system. . . . Copernicus and Galileo believed that a simpler model glorified God, who is 'simple' (theologically, this means not composed of parts)."

Today's geocentrists say the Tychonian theory, which requires fewer epicycles and suchlike than does the Ptolemaic, should have been accepted as the most likely theory, based on Occam's razor. The Tychonian theory had a complexity about equal to that of the Copernican theory, which posited circular rather than elliptical orbits and ended up requiring epicycles of its own. But, says Faulkner, later phenomena, "such as aberration of starlight and trigonometric parallax are better explained in the heliocentric model rather than [in] any geocentric theory." In other words, taking the best geocentric and heliocentric models, such phenomena are more easily explained through heliocentrism.

The Ptolemaic system stood for a millennium and a half, "making it one of the most successful scientific theories of all time." Little by little, though, it came to be seen as inadequate; adjustments to the original, elegant theory had to be made to "save the appearances." This was done chiefly by adding epicycles. "By the Renaissance, the Ptolemaic model had become very unwieldy," says Faulkner, "which led many people, such as Copernicus, to conclude that the model may not be correct. It is not clear if Ptolemy actually intended the theory to be taken as a statement of reality. It could be that he meant it merely as a method of calculating planetary positions. If so, this would have been a very modern view of what a theory is."

It is a view that comes easily to hand for us but one that was foreign to the ancients. They would have expected any theory to be a true reflection of celestial reality. They would not have thought in terms of a theory that was known not to represent reality accurately but that was useful in determining the motion of planets and stars.

At the end of his paper Faulkner summarizes the progress of astronomical thinking. Copernicus (1473-1543) "put forth arguments for the heliocentric theory but also worked out the relative sizes of the orbits and the correct orbital periods of the planets for the first time." He made mistakes—such as thinking that the orbits were circles; they very nearly are, but not quite—but his work represented a substantial advance.

Later, Johannes Kepler (1571-1630) refined the Copernican system by positing that the planets' orbits are actually ellipses with the Sun at one focus of each ellipse. This is the first of Kepler's three laws. His other two laws establish the rates at which the planets move in their orbits (at all times in any planet's orbit, the planet-Sun vector sweeps out the same area per unit time) and a relationship between the periods and sizes of the planets' orbits (the cube of the radius—strictly, the semi-major radius—is proportional to the square of the period).

Kepler empirically, rather than theoretically, deduced his three laws through consideration of the observations that had been carefully made by Tycho Brahe (1546-1601). These observations were much more precise than those available earlier. The earlier ones were sufficient to justify the Ptolemaic theory, but Tycho's finer observations made the old model obsolete. On his own part, Tycho developed the modified Earth-centered model that bears his name, but he was unable to make the refined use of his own measurements that Kepler eventually was able to.

Faulkner says, "Decades after Kepler, Isaac Newton (1643-1727), using his newly discovered calculus and mechanics, was able to deduce Kepler's three laws of planetary motion theoretically. This was taken as a great triumph of Newtonian mechanics and verification of Kepler's work."

BIBLE AS BLUDGEON

TO DANNY FAULKNER'S 10,000-word critique of his views, Gerardus Bouw wrote an 8,000-word reply titled "Geocentrism: A Fable for Educated Man?"[47] In his second sentence Bouw makes a remarkable—and easily refuted—claim: "few non-geocentric creationists have done more than a cursory investigation of geocentricity. Invariably, those who do take more than a cursory look become geocentrists."

Faulkner, who is a creationist, has made "more than a cursory investigation of geocentricity," as proved by his long article about Bouw, yet he has not adopted geocentrism. He works for what may be the largest creationist organization in the country, Answers in Genesis. Have others on its staff made "more than a cursory investigation of geocentricity" and then become geocentrists? Bouw offers no evidence that anyone has made such a move.

His claim is gratuitous and entirely without foundation—unless he argues that anyone who has not become a geocentrist has not yet made an investigation that rises above the

47. Geocentricity.com/ba1/fresp.

cursory, which amount to "If you don't agree with me, that means that you haven't thought things through; everyone who has thought things through agrees with me."

Following his tendentious introduction, Bouw says of Faulkner's article that "it is very shallow and often misrepresents geocentrists, the history of the Copernican Revolution, its evidences, and the authority of Scripture. It fails to deal with any of the hard issues." *Any* of the hard issues? It deals with parallax and star streaming, among others. If these are not consider among the "hard issues," Bouw fails to indicate which topics qualify for that label. If Faulkner's critique of Bouw's arguments is "shallow," Bouw fails to list Faulkner's shallows alongside a list of his own depths.

As an aside, it is relevant to note Bouw's brief mention of Faulkner in Bouw's chief work, *Geocentricity*. In the penultimate chapter, called "Geocentrists and Their Critics," Bouw has a subsection on "Straw Men." This is where Faulkner's name appears. Bouw says that Faulkner and others "are not above erecting straw men and knocking them down." As an example, he points to Faulkner's "claim that the astronauts on the Moon saw the Earth rotating therefore they've proven the Earth rotates is a straw man. If you believe that is a proof, then you also have to believe that while riding on the horse of a carousel and you see the central support rotating; it proves that the central support rotates and that you're not turning about it."[48]

Ignore Bouw's awkward phrasing. He believes he has here a conclusive argument, the conclusiveness of which demonstrates that Faulkner brought up a straw man. Bouw

48. *Geocentricity*, 671.

likens the Earth to the central pillar of a carousel and the Moon to one of the carousel's horses. The rider on the horse and the astronaut on the Moon perceive relative motion, but what moves? Is the Moon orbiting the Earth, or is the Moon virtually still while the Earth rotates on its axis beneath it? Does the rider go around the central pillar, or does he remain motionless as the pillar spins? Bouw rightly infers that the rider circles the pillar and that the pillar is motionless. He makes a parallel to the astronauts' case: the Moon circles the Earth, while the Earth remains motionless. But he forgets something important.

If the Moon circled the Earth every twenty-four hours, and if the astronauts patiently observed that motion, they would see that the Moon's backdrop would change hour by hour. In the course of a day the whole of the star field would pass behind the Moon. What they observed in fact is that the backdrop remains fixed, which only could mean that the Earth rotates on its axis with respect to the star field.

Back to "Geocentrism: A Fable for Educated Man?" Bouw distinguishes his notion of geocentricity from geocentrism: a key point of geocentrism is that the Earth is at the very center of the universe, but geocentricity holds that the static Earth "is actually offset from the geometric center of the universe." Then he notes that, in Galileo's time, "the Tychonic system and the Copernican system were neck and neck in terms of acceptance." He now is one-third of the way into his paper and says, "By now the reader may have noticed that the misstatements and errors in Faulkner's paper are so manifold that it would take a very long paper to counter them all"—but by this point in his discussion Bouw has alleged few errors on Faulkner's part and then only minor ones. He has not demonstrated that Faulkner's errors are common or large enough to warrant the adjective "manifold."

Then come biblical considerations. Bouw keeps returning to the Bible when the emphasis ought to be on scientific demonstration. It appears that his heart lies more with exegesis of the sacred text than with the subject matter of his professional degrees. He refers to himself in the third person: "Bouw relies entirely on the King James Bible and even rejects the authenticity of the Septuagint. Faulkner claims without a thought [does Bouw have any proof for this?] that it is 'the original languages of Scripture that matter, not any translation.' Proof? None is offered. Indeed, there is none. Not a single scripture says that a translation is worse than an original. It does imply the contrary."

Where does it imply that? In a footnote Bouw quotes Hebrews 11:5: "By faith Enoch was translated that he should not see death; and was not found, because God had translated him: for before his translation he had this testimony, that he pleased God" (KJV). This is an egregious misunderstanding on Bouw's part. He tries to argue that a translation of the original biblical texts into a modern language—he has in mind, of course, the King James Version—can be as good as or even better than the original. His sole proof is a verse in which appears the past tense of the verb "translate," which is used not in terms of putting words from one language into another but in terms of movement.

In Hebrew 11:5 there is nothing about rendering words into a foreign tongue. The verse is about Enoch being taken up to heaven ("translated" to heaven). Protestant versions that are more recent than the King James say that

Enoch "was taken up so that he would not see death" (New American Standard) or "was taken from this life" (New International Version).

Bouw sees the word "translate" in Scripture, sees that it has a positive connotation (being taken up to heaven is a good thing), and concludes that verbal translation is like spatial translation. It is as though he equates the Japanese currency, the yen, with a strong desire or craving or as though he thinks the rear storage compartment of an automobile is the proboscis of an elephant because Americans use one word for both things.

Bouw wants to keep to his principle that the King James Version is the most authoritative available, on a par with the original texts. He notes that the Gospel of Matthew generally is understood to have been written in Aramaic or Hebrew. That original text is lost to us; all we have is a Greek translation. Since our basic text is a translation, it "has no more authority than a King James Bible," even though the latter is, at least for the Gospel of Matthew, at best the translation of a translation (from Aramaic or Hebrew to Greek and then to English).

Bouw continues his linguistic remarks. The words "sunrise" and "sunset" seem natural to a geocentric viewpoint: if we say the Sun rises and sets, the implication is that it is doing the moving and that the Earth is standing still. If heliocentrism were true, why didn't we see developed words such as "tosun" and "fromsun," neologisms coined by Bouw? His answer is that "since God founded the languages, and if heliocentrism were the true state of affairs, then it would be a simple matter for God to have created words like 'tosun' and 'fromsun' instead of 'sunrise' and 'sunset' to better encapsulate the 'truth' of heliocentrism."

How does Bouw come to the notion that "God founded the languages"? What does he mean by that phrase? Does he remove all instrumentality from men, giving them no liberty to form their own languages over time? If God "founded" languages, why do languages keep changing (can't God get it right the first time?), why are there so many of them, and why have so many languages disappeared? (No one speaks Hittite today.)

In his very next paragraph Bouw complains that "all Faulkner can do is to ridicule" the geocentric position, but doesn't Bouw invite ridicule when he makes this sort of argument? His logic is this: God, not man, forms languages. God conforms languages to physical reality. If heliocentrism were true, God would have arranged for words such as "tosun" and "fromsun" that would reflect that truth, but the words he gave us were "sunrise" and "sunset." This means the Sun literally rises and sets. Ergo, geocentrism is true.

Next Bouw complains that Faulkner fails to address "the strongest geocentric passages" of Scripture. "Faulkner has no way to refute their geocentric impact." Let's give the verses as Bouw gives them:

Joshua 10:13: "And the Sun stood still, and the Moon stayed, until the people had avenged themselves upon their enemies. Is not this written in the book of Jasher? So the Sun stood still in the midst of heaven, and hasted not to go down about a whole day."

Ecclesiastes 1:5: "The Sun also ariseth, and the Sun goeth down, and hasteth to his place where he arose."

Malachi 4:2: "But unto you that fear my name shall the Sun of righteousness arise with healing in his wings; and ye shall go forth, and grow up as calves of the stall."

Now let's turn to Bouw's exegesis. Joshua 10:13 is thought by almost all geocentrists to be their strongest scriptural passage. Bouw says that "God could have said, 'And the Earth stopped turning so that the Sun appeared to stand still,' but he didn't. In effect, Faulkner claims that since it was inconvenient for God to tell the truth, he promoted the commonly accepted story, although the Holy Ghost knew it not to be true." This is unfair to Faulkner, who nowhere claims that "it was inconvenient for God to tell the truth" and who nowhere says that "the Holy Ghost knew it not to be true."

The Bible is replete with imagery and symbolism. Jesus says, "I am the door" (John 10:9). Not a single Fundamentalist Protestant thinks that he is a slab of wood with hinges on one side and a latch on the other. Would Bouw claim that Jesus "knew it not to be true" when he said he was a door? Was it a matter of it being "inconvenient for [Jesus] to tell the truth"? If Bouw's exegesis of Joshua 10:13 is the best he can muster, then his biblical arguments will fail to impress anyone who is not already both a geocentrist and a Fundamentalist.

Bouw gives a similar argument regarding Ecclesiastes 1:5. "Again, God could just as well have spoken the 'geokinetic truth' by simply adding the sense 'seemeth to' before each of the three actions. That is, to say instead 'the Sun also seemeth to arise, and the Sun seemeth to go down, and seemeth to haste to his place where he arose.' Why did God persist in his geocentric 'error'?"

This argument highlights a basic misunderstanding on Bouw's part regarding the composition of Scripture. He apparently assumes that inspiration implies dictation. "God could just as well have spoken" such-and-so, he says. In truth, in only a few places in Scripture do we have evidence of dictation by God. In the Old Testament, those places are found chiefly in the Pentateuch and in some of the prophets; in the New Testament they are found solely in Revelation.

In the rest of the sacred text we see the sacred writers writing according to their best lights—under the inspiration of the Holy Spirit, yes, but not as amanuenses and not as automatons. In most cases there is no indication whatsoever that the sacred writers, at the time they were writing, understood themselves to be guided in some manner by God. Inspiration is more like a gentle nudge than a sharp rap on the knuckles. If God can write straight with crooked lines, he can get his points across in non-coercive ways.

Of the three verses that Bouw lists as "the strongest geocentric scriptures," his last seems oddly chosen. "Now note Malachi 4:2, where the Sun, as a type of Jesus (see also Psalm 19:1-6), is said to rise. It is clear that this refers to the Resurrection. How, then, can a believer in the Resurrection of the Lord Jesus Christ insist that the word 'arise' is literal truth when referring to the Resurrection here, yet at the same time insist that it is not literally true when applied to the Sun here, in this same verse?" The answer is that "arise" is not really being applied to the Sun at all. In this passage "Sun" does not refer to the star but to Christ, who will "arise with healing in his wings." The imagery of wings can be applied to Christ if he is likened to the light-giving

Sun but not to the Sun itself when considered as a datum of astronomical investigation.

A few paragraphs later Bouw returns to his second key verse, saying,

> So, if Genesis 1:1, "In the beginning God created the heaven and the Earth" is a clear statement that God created, then Ecclesiastes 1:5, "The Sun also ariseth, and the Sun goeth down, and hasteth to his place where he arose," is just as clear a statement of geocentricity. And with that, we come to the *real* issue: Is the Scripture to be the final authority on all matters on which it touches, or are scholars to be the ultimate authority? The central issue is not the motion of the Earth, nor is it the creation of the Earth. The issue is final authority; is it to be the words of God or the words of men?

The words of God—as interpreted by whom? Bouw holds to a position of private interpretation, believing the Bible to be clear and unambiguous to those who have real faith. Like other Fundamentalists, he believes he has grasped the only accurate understanding of each verse. This is not the place to disabuse him of that notion, but it is the place to point out that, throughout his critique of Faulkner's critique, Bouw falls back on scriptural interpretations that are peculiar to a small subset of Christians and, sometimes, that seem to be peculiar to himself alone. If Faulkner doesn't always seem to have a convincing answer (and sometimes he doesn't), at least he avoids trying to forge the sacred text into a weapon on behalf of his scientific theories.

After looking at "the real geocentric scriptures that Faulkner cannot refute," Bouw briefly turns to "historical issues." The only one he really examines is "the possibility that Tycho Brahe was poisoned." This is a favorite trope for geocentrists who want to discredit Tycho's assistant, Johannes Kepler. It is Kepler on whom they throw suspicion, though it is unclear what Kepler could have hoped to gain from murdering Tycho—he already had access to Tycho's data—but Kepler was a heliocentrist and therefore not trustworthy.

Bouw cites 1996 forensic work done by the University of Lund in Stockholm on some hairs taken from Tycho's body; traces of mercury were found. "Of course, that is no proof that Brahe was poisoned by someone else, but it does beg the question of why he would be so careless that one time when the rest of the hairs showed no lethal abundance, even given that he routinely worked with mercury and arsenic." After this confusingly-worded sentence, Bouw says, "It really would help the cause of truth if Faulkner had done his homework instead of making rash and unfounded charges and innuendos"—such as the one Bouw just made, one presumes.

As it turns out, his suggestion has proved groundless. In 2012 a new investigation, this time by Czech and Danish scientists, tested bone, hair, and clothing taken from Tycho. Mercury was found in traces far too small to substantiate murder, and there was no lethal level of any poison. The team concluded that "it was impossible that Tycho Brahe could have been murdered."[49]

In the conclusion of his paper, Bouw says, "In examining Faulkner's case against geocentricity we found that his insistence that the Scriptures do not present a geocentric universe is not founded on any reason other than his opinion." Faulkner, if he chose, could respond with an equally gratuitous remark—"Bouw's case is not founded on any reason other than his opinion"—but he does not stoop to that, though it is truer of Bouw than of Faulkner.

Granted, Faulkner doesn't examine every claim made by Bouw in *Geocentricity*; on the other hand, Bouw does not grapple with every point Faulkner makes in his short paper. The two largely talk past one another, and the reason for that is that Faulkner sticks mainly to questions of science while Bouw sticks mainly to questions of exegesis.

Faulkner, as a creationist, worries that geocentrists "offer a very easy target of criticism for our critics." He worries that the small number of creationists who are geocentrists will bring discredit to the entire movement. Bouw calls this "sheer nonsense. Evolutionists, atheists, and agnostics in the know can easily shame creationists on the issue of geocentricity by simply pointing out the hypocrisy of their insistence that the days in Genesis 1 are literal while the rising and setting of the sun is [sic] not."

He has a point here, and it is one that unimaginative anti-creationists (whether unbelievers or believers) might use, but it is not much of a point. Most people will accept

49. Newsfeed.time.com/2012/11/17/was-tycho-brahe-poisoned-according-to-new-evidence-probably-not.

that, in interpreting one portion of Scripture literally, one is not obligated to interpret all sections literally. (Even Bouw doesn't do this in practice.) There is no hypocrisy in this; there is only an attempt to interpret differing passages in differing ways, according to their styles: history as history, poetry as poetry, imagery as imagery. That many "evolutionists, atheists, and agnostics" don't understand this—never having bothered to consider the various senses of Scripture and working from a prejudice that tells them that they don't need to do so—is not the measure, or at least it should not be.

Bouw's penultimate sentence is ungracious: "We conclude that the creationist's desire to reject [geocentrism] can only be for the sole purpose of appearing intellectual and acceptable to the world." Bouw is incapable of imagining that fellow creationists might disagree with him because they find his science unconvincing and his exegesis puerile.

THE STORY THUS FAR 3

GERARDUS BOUW IS one of the few geocentrists with a doctorate in astronomy. It is curious, then, that much of his pro-geocentrism argument is couched in biblical terms. As Robert Sungenis later would do, Bouw wrote a long work in defense of geocentrism and then, finding few people willing to wade through it, wrote a much shorter treatment.

Bouw thinks that the Copernican revolution "was not just a revolution in astronomy, but it also spread into politics and theology." For Bouw and for others, problems in politics and theology have stemmed from a problem in cosmology. A wrong understanding of science has led to a wrong understanding of Scripture, which in turn has led to errors in other areas of life. (Note that Solange Hertz thinks the sequence was reversed, with political and theological errors preceding the acceptance of heliocentrism.)

In his major work, *Geocentricity*, Bouw provides biographical sketches of prominent geocentrists of the last half century, though the book mentions only one Catholic. The emphasis is on Protestants whom Bouw has known or has learned from. They and Bouw are a minority within a minority: most other Protestant creationists disagree with them on whether the Earth orbits the Sun or the Sun the Earth. One such naysayer is Danny Faulkner, who is associated with a large creationist ministry and who, like Bouw, has a doctorate in astronomy. Faulkner faults Bouw's science and his exegesis. Tellingly, when Bouw answers, it is mainly in terms of exegesis.

This is common among geocentrists. When met with scientific arguments that are difficult to answer, they respond with scriptural citations. Sometimes they just ignore the scientific challenges.

PART 4

COMMITTING LITERARY BURGLARY

A T GEOCENTRISMDEBUNKED.ORG IS an article titled "Top Geocentrists Caught Plagiarizing." The writer is David Palm, the website's owner and a long-time opponent of geocentrism. By profession an engineer, Palm has a degree in New Testament studies from Trinity Evangelical Divinity School. He is a convert to the Catholic faith. Radio host Al Kresta has termed him "an astute lay theologian," but Palm eschews the title. He says he is just a layman deeply interested in Catholic beliefs and practices.

When interviewed by Kresta, Palm claimed the new geocentrists' errors about science, Church history, and scriptural interpretation "draw people into a dark and conspiratorial view of the world." While trying to prove their points, he said, some of the new geocentrists have stooped to plagiarism. In his article, Palm provides two definitions.

Westminster Theological Seminary's "Statement on Plagiarism" says "Plagiarism is literary burglary":

> At its worst it involves an outright intent to deceive, to pass off another's work as one's own. More often,

it is the result of carelessness or ignorance. But whether intentional or unintentional (the distinction is often hard to draw), plagiarism is always an error, and a serious one. Whenever you borrow another writer's words or ideas, you must acknowledge the borrowing. . . . When you use their words, their ideas, even their organization or sequence of ideas, say so—in a footnote or in the text. Claim as your own only what properly is yours [Westminster Theological Seminary, "Statement on Plagiarism"].

What is plagiarism? . . . Intentionally representing the words, ideas, or sequence of ideas of another as one's own in any academic exercise; failure to attribute any of the following: quotations, paraphrases, or borrowed information [George Washington University, "Citing Responsibly"].

Palm says that the "new geocentrists present themselves as both qualified and competent to overturn the entire world of physics and astrophysics concerning the motion of the Earth. They also present themselves as uniquely trustworthy and honest in their dealings with the scientific data." But are they? Geocentrists find themselves arrayed against virtually the whole of the scientific community, which includes not just many unbelievers but many devout Christians. According to geocentrists, says Palm,

[the] vast majority of working scientists . . . have misrepresented the physical evidence of the

universe, often intentionally, and precisely just in order to deny geocentrism (a cosmology which these scientists allegedly know in their heart of hearts to be true). Robert Sungenis demonstrates this dark, conspiratorial view of the scientific community in the title of his talk at a 2010 geocentrism conference, "Geocentrism: They Know It But They're Hiding It."

Palm adds that Sungenis reveals this mindset "even in his movie, *The Principle*," in which "he accuses NASA of intentionally removing information from its websites that would support geocentrism." In fact, Sungenis thought this point so important that he included his passing comment about NASA in the trailer to the film.

"As the geocentrists spin it," says Palm,

they alone have the competence, courage, and honesty to present things as they really are. The problem is that the geocentrists have repeatedly demonstrated both incompetence and unethical behavior. . . . They have presented themselves as the lone, trustworthy sources of information in this area and are striving to gain as many followers and donors as they can.

This is, obviously, serious business. Therefore, it is legitimate and important to examine their behavior, so that those who are deciding whether or not to follow them can more accurately answer these central questions: are the geocentrists competent, and are they trustworthy?

In "Top Geocentrists Caught Plagiarizing" Palm looks at the work of three geocentrists: Sungenis, Robert Bennett, and Gerardus Bouw. First we will consider his charges against Bouw, who, until recent years, deserved to be considered the most prominent geocentrist and still may be so among his fellow Protestants.

"The most egregious example of plagiarism amongst the geocentrists (at least so far) is found in a recent article[50] by Dr. Gerardus Bouw," says Palm. The article concerns the Lagrange points and a solution to the so-called "three-body problem." The Lagrange points are the five points in space at which a small object, such as a satellite, can maintain a stable orbit with respect to two large bodies, such as the Earth and the Moon, maintaining the same period of rotation as the large bodies.

At those points the gravitational pull of the large bodies and the acceleration of the small object balance out in such a way that the latter can remain virtually stationary with respect to the former. Every other placement for the small object results in an unstable orbit for it, with it eventually crashing into one of the other bodies or finding itself flung into outer space or moving into an orbit with a different period.

Bouw's article was written in response to a challenge issued by astrophysicist Tom Bridgman in 2010. Bridgman

50. Gerardus Bouw, "A Geocentric Solution to the Three-Body Problem" (2014), galileowaswrong.com/wp-content/uploads/2014/08/Geocentricsolutionto3-bodyprob.pdf.

noted that it is not particularly difficult to account for the Lagrange points—both for their existence and their locations—in a heliocentric system, but no one had yet shown how to account for them in a geocentric system. Thus his challenge, which went unanswered for four years until Bouw's 2014 article appeared.

Bouw explains that "some humanistic, self-professed scientists proposed that, for the three-body problem, the theory of geocentricity should yield a different solution than what is observed. In anticipation of that challenge Bob Sungenis invited me to derive a geocentric solution within the framework of geocentricity. Our purpose was to derive a viable geocentric framework for the three-body problem. We were successful, and this paper is the fruit of our labors."

Palm reports that "Dr. Alec MacAndrew, in his analysis of Bouw's article, notes that the first section contains outright mathematical errors and the third section is a verbatim copy (albeit referenced) of a significant portion" of an article by Neil J. Cornish. "While the mathematical errors certainly call into question Dr. Bouw's basic competence," says Palm,

his copy-pasting almost the entirety of somebody else's work is ethically dubious, at best, and is made worse by the fact that Bouw sandwiches Cornish's material between introductory and concluding text of his own without any distinction. "The first paragraph of his section on page 26 and the two concluding paragraphs on page 30 are written by Bouw [says MacAndrew], and everything in between is an exact copy of Cornish's paper. Bouw does nothing to distinguish his words and Cornish's—no quotation marks, no italics, no indentation. The fact that he gives a reference in this case is hardly sufficient to avoid plagiarism."

MacAndrew holds three degrees in physics from Imperial College, London, including a Ph.D. awarded in 1982. He worked in various industrial and consulting capacities, now is retired, and "has the opportunity to pursue science topics" that interest him. One of those is geocentrism. He has contributed several articles to GeocentrismDebunked.org.

Palm says "it is the second section of Bouw's article that most interests us here. Under a heading 'Equations of Motion for the Infinitesimal Body,' Bouw proceeds to copy an entire eighteen pages of material from F. R. Moulton's *An Introduction to Celestial Mechanics* (pages 277-294), without any attribution at all." It would be one thing, says Palm, if Bouw simply had copied Moulton's work without attribution, but "he goes farther than that." Palm quotes MacAndrew:

Bouw copies the entire eighteen pages. He copies the mathematical expressions *exactly* using identical notation to the original. In some cases he paraphrases the text between the mathematical expressions, in others he copies it word for word, but in every case the import of his text is identical to the original. . . . Bouw also copies the diagrams from

Moulton's book exactly—they appear to be scanned from the original. Amazingly, Bouw has gone to the trouble, not just here but with the Cornish paper . . . of transcribing all the many complex equations using LaTeX or some other equation editor in identical form to the original. That labour of deception must have been dismally tedious. [I have preserved MacAndrew's British spelling.]

MacAndrew asks, "What is going on here? Did Bouw really expect people to believe that he had developed this derivation himself? Could he possibly think that? He must have done, for he introduces this section by writing, 'There remains for *us* to derive the generalized equations of motion describing the path of the test mass, M. Then *we* need to solve the generalized equations for locations where M's velocity is zero relative to both Sun and Earth'" [emphasis added by MacAndrew]. MacAndrew continues:

It takes only a cursory inspection to see that the diagrams are in an outmoded style, that the mathematical notation is old-fashioned, and that there are technical terms that few would use today (for example, talking about the 'right member' of an equation where most today would say the 'right-hand side,' an anachronism that Bouw repeats). It is immediately obvious that this is copied from an older source, and it didn't take me long to find Moulton's original book in Google Books and to find a complete copy of the book online in the Open Library.

MacAndrew concludes this way:
This is a double failure for Bouw and for geocentrism. Their so-called "successful" derivation of a "viable geocentric framework for the three-body problem" turns out to be nothing more than the dishonest counterfeit of a classical Newtonian derivation which is based on equations of motion in which the Earth and the Sun orbit their mutual centre of mass (and so which is clearly not geocentric). Not only does the derivation not do what he claims for it, but his reputation for honesty, such as it was, is in tatters.

Before we turn to Palm's discussion of plagiarism by Robert Bennett, it is useful to provide here his conclusion, which begins with a quotation from Tom Bridgman: "Everything I've seen from geocentrists is a cheat, trying to take someone else's heliocentric solution and then moving the origin to the Earth."[51] Palm believes his article supports Bridgman's contention, and he asks his readers to ask themselves these questions:

Why do these men, who are unable to make their own case, through their own efforts, feel qualified

51. Tom Bridgman, "Failing More Basic Physics" (2011), dealingwithcreationisminastronomy.blogspot.com/2011/06/geocentrism-failing-more-basic-physics.html.

to completely overturn a broad scientific consensus and insist that we listen to them instead? What does it say about their competence that they have to co-opt, modify, and then present as their own the hard work done by others? And what does it say about the image they have tried so hard to cultivate of themselves as uniquely honest and operating with the highest integrity?

NO COPYRIGHT, NO FOUL

D AVID PALM'S ARTICLE "Top Geocentrists Caught Plagiarizing" is about 4,300 words long. The reply by Robert Sungenis contains over 19,700 words. Its title is "David Palm Caught Falsely Accusing Opponents of Plagiarism."[52] It was uploaded to Sungenis's website, GalileoWasWrong.com, on August 28, 2014, just six days after Palm's article appeared. Here we will look only at those portions of Sungenis's paper related to Palm's charges against Gerardus Bouw.

Sungenis describes Bouw as "a Protestant Christian, and I've known him for about ten years"—that is, from not long after Sungenis began to write in favor of geocentrism. Bouw sent to Sungenis "two simple paragraphs, which answered Mr. Palm's charges quite succinctly." In an e-mail dated August 25, 2014, Bouw wrote:

---

52. Robert Sungnis, "David Palm Caught Falsely Accusing Opponents of Plagiarism" (2014), debunkingdavidpalm.com/ddp/docs/David%20 Palm%20Caught%20Falsely%20Accusing%20Opponents%20of%20 Plagiarism.pdf.

This is one of the problems that arises when an incomplete work is published. I made no secret of using Moulton; both [James] Hanson and Frank Wolff know that I was doing so. I certainly did credit Cornish and even credited Goodman, whom our critic does not mention. I even gave the link to their paper. As far as I know, Forrest Moulton's work has not been reprinted, nor his copyright renewed, since the edition I used, which was released in 1914.

Palm did not get my permission to use my picture, neither did the source website obtain that permission. A slight touch of hypocrisy?

Let's deal with Bouw's last paragraph first. It is true that Palm didn't ask his permission for the use of his photograph, but the "source website" certainly did. That website is the one for Baldwin Wallace University, which is located in a suburb of Cleveland. The website lists Bouw as emeritus professor of computer science—right next to the photograph used by Palm. If Palm ought to have asked for permission, he should have asked the university, not Bouw. So, no, there was no "slight touch of hypocrisy" on Palm's part. Now let's turn to the substance of Bouw's first paragraph.

Sungenis asks, "Why didn't either Palm or MacAndrew bother to look up whether Moulton's 1914 copyright was renewed?" Sungenis answers himself: "Because both these vigilantes got too wrapped up in making up their own rules concerning what constitutes plagiarism due to the fact that they have an agenda to discredit their opponents since they can't beat their opponents on scientific grounds."

Note first Sungenis's *ad hominem* ("these vigilantes")— an argumentative method he uses often—and then note that he argues in terms of copyright. Apparently he thinks that lifting material from a work the copyright of which has expired is okay and that plagiarism consists only in lifting material, without proper attribution, from works still under copyright. At this point it would be good for the reader to go back a few pages and look at the definitions of plagiarism as given by George Washington University and Westminster Theological Seminary. Sungenis received his bachelor's degree from the first school and his master's degree from the second. Neither school defines plagiarism in terms of copyright—nor does any other university. (Nor does any dictionary.)

The works of Charles Dickens are long out of copyright. It would be plagiarism to incorporate chunks of one of his novels into one's own writing, unless those chunks were surrounded by quotation marks and properly referenced. The same rule holds for the text of a novel published this year. That new novel would be under copyright, and Dickens' novel would not be—but that makes no difference whatsoever. Sungenis and Bouw try to distract the reader by bringing up the issue of copyright.

With respect to Bouw's alleged plagiarism, Sungenis attempts no refutation of Palm's (and MacAndrew's) claims. The best he can do, at the end of his section about Bouw, is to quote Bouw's e-mail: "This is one of the problems when an incomplete work is published." Sungenis says, "In other words, the paper is not complete. Dr. Bouw told me that he is still working on the transform equations and some other material that will be added to the present article."

Those things are irrelevant to the question whether Bouw's article, as made public in its current (supposedly incomplete) state, contains plagiarism. If so, then what Bouw ought to add first, before emendations "on the transform equations and some other material," are quotation marks and footnotes.

Sungenis's response doesn't address the problem MacAndrew identified: that Bouw copied eighteen pages from F. R. Moulton's book, keeping the mathematical expressions exactly as Moulton had them but "paraphras[ing] the text between the mathematical expressions." This could not have been a mere oversight. It was not a matter of forgetting to put quotation marks around a long passage that had been copied and pasted. That could be an explanation if the offending text had been left intact, but Bouw edited the text, yet not enough to eliminate Moulton's archaic expressions, such as referring to the "right member" of an equation. There is no obvious way for Sungenis to defend how these eighteen pages were handled, so he ignores them and brings up copyright instead, but copyright is a non-issue. It has nothing to do with whether plagiarism occurred.

QUESTIONS OF DEGREES

ROBERT BENNETT IS a fairly new convert to geo-centrism. He came to it late in life, around 2004. He was in his mid-sixties when he read an essay by Robert Sungenis at the Catholic Apologetics International website. The essay led him to consider what Scripture had to say about geocentricity. He concluded it said enough to warrant belief in the idea. There followed not just a friend-ship with Sungenis but a cooperation in the latter's publish-ing and even in his education. The publishing consisted of contributing the second-longest and most technical chapter to *Galileo Was Wrong: The Church Was Right*. As controver-sial as that contribution has been, Bennett's contribution to Sungenis's education has been more so.

Sungenis claims two legitimate degrees, a bachelor's from George Washington University and a master's from Westminster Theological Seminary. He has given conflict-ing accounts of what his undergraduate studies focused on.

When asked about his knowledge of science, Sungenis wrote, "I've been studying science all my life. I was a chem-istry and physics major in college, studying for pre-med.

I'm an avid reader of *Scientific American*, *New Scientist*, *Nature*, and about a half-dozen other science magazines on a weekly basis. I own a library of science books."[53] Note that he claims to have had two majors at George Washington University: chemistry and physics. The implication is that he received his undergraduate degree in both subjects. This is bolstered by an article that appeared at the website of the Kolbe Center for the Study of Creation, a Catholic creationist organization. There Sungenis wrote, "I was a chemistry and physics major in college."[54]

This does not square with what he wrote to me on February 2, 2003: "One of my majors in college was physics, with an emphasis in astronomy. I also majored in religion. That should also answer the question about 'formal' training, since one doesn't get any more 'formal' training other than the course work the university offers."[55] In this telling, he had double majors, but they were physics and religion, not physics and chemistry. The implication still was that he was awarded a degree that reflected a double major.

The next modification of his background came on June 18, 2013, when Sungenis published at his website a long article called "Anatomy of a Smear."[56] A third of the way through it he wrote, "I have said many times I was a physics major in college." Then he added a caveat: "For the record, I don't need a 'degree' in physics in order to know

53. Doxacommunications.com/sungenis/discussion-with-a-modern-priest-about-the-interpretation-of-scripture.
54. Kolbecenter.org/the-case-against-theistic-evolution.
55. E-mail to the author (Feb. 2, 2003).
56. Robert Sungenis, "Anatomy of a Smear" (June 18, 2003), Galileowaswrong.com/wp-content/uploads/2014/06/Anatomy_of_a_Smear_Campaign-1.pdf.

the physics. In today's scientism (as opposed to real science), a degree in physics often means that you have been indoctrinated by the reigning powers of atheism."

That he majored in physics is reiterated on the "About the Authors" page that appears after the table of contents in the first volume of *Galileo Was Wrong*. In the tenth edition, published in 2014, Sungenis says that he "holds advanced degrees in theology and religious studies and was a physics major in college." There is no mention of a bachelor's degree in religion—the implication is that his undergraduate degree was in physics only.

The story changed later in 2014, when, in reply to a online post I wrote, Sungenis said: "[Keating] knows that I have said many times I was a physics major in college. . . . I was a physics major in college before I switched to religion."[57] So here is an admission that his undergraduate degree was solely in religion. It was not in chemistry or physics. So far as physics goes, Sungenis never has shown that he took any courses in that subject beyond the one or possibly two required of all undergraduates at George Washington University.

As for his minor, Sungenis wrote, "I was a psychology minor in college. The day I dropped out of the program was when the professor gave us the results of a comprehensive study of the three major psychologies: Freudian psychoanalysis, Skinner's behavioral system, and the Humanist-Existential."[58] He does not say what his new minor was.

Sungenis's claims about his undergraduate studies have

57. Ibid.
58. Robert Sungenis, "Flunking the Test," *Culture Wars* (Jan. 2010), 43.

been contradictory, but they have not caused him as much grief as have questions about his claimed Ph.D. in religious studies. That degree was granted by Calamus International University, which widely is considered to be a diploma mill. Sungenis rejects that label, saying that Calamus, which is based on the South Pacific island of Vanuatu, chose not to seek accreditation from any of the standard American accreditation agencies because it chiefly caters to students outside the U.S. He says that Calamus

> is just as rigorous as many other U.S. and E.U. accredited degrees in that it requires at least two to three years of intensive research in one's area of expertise and must include the supervision of both a professional academic advisor (mine was Dr. Robert Bennett, Ph.D. in physics) and the academic dean (mine was Dr. Morris Berg), which must then be finalized by the writing of a Ph.D.-style dissertation."[59]

The state of Texas disagrees with Sungenis about the status of Calamus. It lists the school as one of about four hundred "institutions whose degrees are illegal to use in Texas." The state penal code says it is a misdemeanor to use such "fraudulent or substandard degrees" in "a written or oral advertisement or other promotion of a business or . . . in employment or in the practice of a trade, profession, or

59. Magisterialfundies.blogspot.com/2014/01/challenge-to-mark-shea-during-his.html?showComment=1391123094355%20\l%20 c6180327450340433232.

occupation."[60] One can obtain a Ph.D. from Calamus for about $2,000.[61] All studies are done on a distance-learning basis, and there is no attendance requirement because the school's campus is non-existent. To the lawmakers of Texas, Calamus qualifies as a diploma mill.[62]

The school's academic dean, Morris Berg, was one of Sungenis's supervisors. Berg claims numerous degrees, his chief being a doctorate in hypnotherapy from an American institution that he does not name. He runs a hypnosis clinic in Surrey, south of London, and deals with "past life healing and associated therapies"—that is, he believes in reincarnation. There is nothing to indicate that he is qualified to evaluate dissertations that deal with either science or religion, and Sungenis's dissertation dealt with both. His dissertation became *Galileo Was Wrong*. Sungenis's other supervisor was Robert Bennett, co-author with him of that book.

Bennett is not prominent in the geocentrist movement. To the extent he is known at all, it is for his long, technical chapter in *Galileo Was Wrong*. In the second volume of the tenth edition, that chapter is titled "Technical and Summary Analysis of Geocentrism." It is divided into three parts: "Does the Earth Rotate?", "Does the Earth Revolve Around the Sun?", and "Does the Sun-Earth System Move?" Those three parts have a total of sixty subsections, averaging about four pages apiece.

60. Thecb.state.tx.us/index.cfm?objectid=EF4C3C3B-EB44-4381-6673F760B3946FBB.
61. Unicalamus.org/tuition.htm.
62. Michigan has a similar list: michigan.gov/documents/Non-accreditedSchools_78090_7.pdf.

David Palm, in his Internet article "Top Geocentrists Caught Plagiarizing," asserts that Bennett "uses the work of others without giving them credit. . . . [T]hey are not just examples of cutting and pasting while perhaps just being sloppy and forgetting to include a reference. Bennett has actually altered the wording somewhat in order to present the ideas and work as his own." Palm says that "even a run-of-the-mill online plagiarism scanner commonly used [to check the writings of] high school and college students easily discovered the plagiarism in *Galileo Was Wrong*."

Palm provides several pages of side-by-side quotations that, he says, demonstrates Bennett's taking material from uncredited sources. For example, Bennett writes about the Foucault pendulum, the motion of which, say heliocentrists, demonstrates that the Earth rotates on its axis. Geocentrists deny that conclusion and say it is not the Earth that is rotating under the pendulum but the stars and planets that are orbiting the Earth and are acting on the pendulum, thus giving the same appearances. Here is how Bennett introduces the Foucault pendulum:

> Conceived as an experiment to demonstrate the rotation of the Earth; the motion of the Foucault pendulum is a result of the Coriolis effect. It must be long and free to swing in any vertical plane. The first Foucault pendulum exhibited to the public was in 1851 of [sic] the Paris Observatory. It was the first dynamical proof of the rotation in an easy-to-see experiment.

Here is text taken from the website of the geophysics department of Ludwig-Maximilians Universität München:

> ... conceived as an experiment to demonstrate the rotation of the Earth; its action is a result of the Coriolis effect. It is a tall pendulum free to oscillate in any vertical plane .... The first Foucault pendulum exhibited to the public was in February 1851 in the Meridian Room of the Paris Observatory. ... Foucault's pendulum was the first dynamical proof of the rotation in an easy-to-see experiment.

Notice that the misused "of" in Bennett's quotation arises because he dropped the phrase "in the Meridian Room" that appeared in the original text. Other wording has been modified slightly: "the motion of the Foucault pendulum is a result of the Coriolis effect" becomes "its action is a result of the Coriolis effect." "It is a tall pendulum free to oscillate" becomes "It must be long and free to swing." "Foucault's pendulum was the first dynamical proof" becomes "It was the first dynamical proof." These changes are incompatible with a plea that the text was cut and pasted inadvertently.

The wording as given by Bennett has no reference, either by footnote or within the body of the text, to the writing at the German university. So it is with other examples given by Palm, who says that, at the time of writing, he was able to make only a cursory investigation of Bennett's chapter yet still found multiple examples of plagiarism.

As mentioned earlier, in response to Palm's article "Top Geocentrists Caught Plagiarizing," Robert Sungenis wrote a reply called "David Palm Caught Falsely Accusing Opponents of Plagiarism." The reply was almost five times as long as the original article. The short section defending Bennett was written not by Sungenis but by Bennett himself. He begins by saying, "How flattering that Mr. Palm rates me among the top geocentrists! In turn I would rate him among the top misinformed nitpickers on the Internet."

Bennett notes that *Galileo Was Wrong* is a long work that "contains over 1,000 footnotes and references. [Actually, it contains 3,135 footnotes.] Even so it's true that in the rush to meet the submission deadline some citations were lost." Perhaps more than "some" in Bennett's contribution to the book. His 231-page chapter, which is more than ten percent of the whole work, is dense with names, dates, and formulas, yet it has only 55 footnotes, less than two percent of the work's total, even though it by far is the most complex chapter.

Perhaps surprisingly, Bennett does not deny lifting text from the German university website. His justification is that the material was not original to that site. "As stated at the German site: 'This article is licensed under the GNU Free Documentation License. It uses material from the Wikipedia article "Foucault pendulum."' . . . Wikipedia was acknowledged as a source in the *Galileo Was Wrong* reference section."

There is no such section in the tenth edition of *Galileo Was Wrong*. No reference section or bibliography is found at the conclusion of Bennett's chapter. At the end of the third volume is a lengthy bibliography, which is followed by a "webliography." The section in the "webliography" dealing with the Foucault pendulum lists three web references, none of them to Wikipedia or to the German university.

Bennett, in his defense of himself, makes no reference to the other examples of plagiarism uncovered by Palm. Instead, he turns the discussion to the definition of plagiarism, saying, "The plagiarism rules for scientific research varies [sic] from the guidebook for literature in general." Even if that were true, it does not follow that in scientific writing it is permissible to omit most citations and to use, as one's own, words that have been taken from elsewhere.

But such taking actually is permissible, says Bennett, who appeals to the fair use doctrine. "The amount of quoted material that is considered 'fair use' varies with the length of the original work and the demands of the copyright holder. . . . None of the quotes I used exceeded 200 words." This is irrelevant: the "quotes" in question were not set out as quotations at all. They were set out as Bennett's own words, and they were modified to distinguish them from the originals. It was not a matter of the fair use doctrine, which deals with how much of a third party's work can be quoted without first obtaining that party's permission. Besides, fair use means throwing quotation marks around borrowed words, and this is precisely what Bennett did not do. Fair use also means not altering borrowed words, and that is something Bennett clearly did.

David Palm wrote a reply to Robert Sungenis's reply: "Madness: The Geocentric 'Defense' on Plagiarism."[63] In a section about Bennett he says it is "Bennett's defense of his

63. David Palm, "Madness: The Geocentric 'Defense' on Plagiarism" (2014), geocentrismdebunked.org/defense-against-plagiarism-madness.

own plagiarism that runs closest to outright madness. Near the start [of his defense] he declares quite boldly what he was doing":

> The target of [*Galileo Was Wrong*] is the general reader, naïve to specifics like these topics. In order to introduce these topics to said general reader, I used the most succinct quotes found on the Internet and edited them for relevant material and simplified some technical phrasing.

Palm's response: "And basically you can stop right there. Yes, Dr. Bennett, that's exactly what you did. And when you do that but don't cite your sources, it's plagiarism. . . . The rules for scientific publishing are the same as for any other kind of publishing—you may not present somebody else's work as if it were your own." Palm notes that Bennett, in an attempt to exculpate himself, cites a definition from the Health and Human Services Department's Office of Research Integrity. The definition says plagiarism is "the unattributed verbatim or nearly verbatim copying of sentences and paragraphs which materially mislead the ordinary reader regarding the contributions of the author." Palm says,

> The examples I cited from Bennett were "unattributed verbatim or nearly verbatim," were "sentences and paragraphs" in length, and misled the reader as to Bennett's contribution since he presented the material as his own. So, according to his own source, he plagiarized.

THE STORY THUS FAR 4

T HE NEW GEOCENTRISTS gain followers by gaining their trust. They present themselves as authorities in multiple disciplines, from physics and astronomy to theology and scriptural exegesis. What their books lack in style they make up in volume: innumerable equations, quotations, and assertions. Many of the assertions are that mainstream scientists—including some of the most famous—have offered up incorrect conclusions through incompetence or—much worse—through a deliberate intention to deceive. By pointing out the supposed errors of their opponents, leading geocentrists seek to raise their own status. As David Palm puts it, they "have presented themselves as the lone, trustworthy sources of information in this area and are striving to gain as many followers and donors as they can."

Some geocentrists, desperate to appear scientifically competent, have fallen into plagiarism. Palm and physicist Alec MacAndrew show that Gerardus Bouw and Robert Bennett have taken liberties with other writers' words, liberties that can't be waved away as mere "cut-and-paste" errors or proofreading oversights. Robert Sungenis, in defending his colleagues, admits that other writers' words have been used without proper attribution—but he excuses that because the words came from books that are out of copyright.

This is not the way scholars go about bolstering their reputations.

PART 5

HIDING AN ANTI-JEWISH PAST

J UST AS A man can be judged by the friends he keeps, so he can be judged by the ideas he keeps. We permit leeway with friendships: we make allowances for someone who has unsavory friends because, for all we know, he may be trying to help the wayward, or he may be unaware how wayward they are. We make less allowance for someone with unsavory ideas, because the ideas he holds tend to reflect his character.

In "Top Geocentrists Caught Plagiarizing," David Palm asserts that "Robert Sungenis has a long and well-documented history of plagiarism, starting in 2002 with his essay 'Conversion of the Jews Not Necessary? The Apocalyptic Ramifications of a Novel Teaching.'" In that essay—which came at the beginning of his decade-long fascination with supposed Jewish conspiracies—Sungenis "plagiarized from a number of sources," says Palm, "including Nazi Labor Minister Robert Ley and almost 1,500 words from white supremacist Gordon 'Jack' Mohr. . . . He has also deployed numerous bogus quotes without documentation (a combination of calumny and plagiarism), including a blatantly perverted 'quote' from Albert Einstein, making it appear that Einstein considered the charge of 'anti-Semitism' to be nothing more than a mere ploy designed to cow Gentiles."

Objecting to this characterization, Sungenis said, "Mr. Palm knows full well that I have publicly stated that I will no longer be addressing any political issues, especially concerning the Jews, and I have made my new direction public knowledge." (In an e-mail to Palm, demanding a stop to online comments about Sungenis's anti-Jewish articles and his endorsement of conspiracy theories, Sungenis threatened: "I could make you look very bad if I really wanted to, and that is only after a cursory review of the Internet and my limited experience with you."[64] In the same e-mail Sungenis complained that Palm mentioned publicly that Sungenis "promoted the belief that the tsunami that led to the Fukushima reactor disaster was caused by nuclear weapons detonated by the Israelis off the coast of Japan, to punish that country for giving nuclear technology to the Iranians" and that Sungenis admitted to journalist Jared Olar that he thought that low-yield nuclear weapons were used to bring down the Twin Towers.)

In a 2014 online exchange Sungenis said, "I publicly disavow any 'Jewish conspiracy theories' I have entertained in the past, and I apologize for anything I have said about the Jews that isn't true. I further say for the record that I am not an 'anti-Semite' and I am not a 'Holocaust denier,' and any statement I have made in the past that might have implied those positions, I likewise disavow."[65]

64. E-mail from Robert Sungenis to David Palm (Dec. 19, 2013).
65. Robert Sungenis, "Correcting the Misinformation of David Palm" (July 23, 2014), debunkingdavidpalm.com/ddp/docs/Correcting%20the%20 Misinformation%20of%20David%20Palm.pdf.

These comments were carefully phrased. Sungenis did not stipulate which of the things he "said about the Jews" may have been untrue. The implication is that, while he apologized for untrue things, he still affirms those things about Jews that he believes to be true. But which beliefs of his fall into which category? On that he has been silent. Then there is his curious formulation that he "will no longer be addressing" Jewish topics and the admission that

> I still have the same beliefs. That will never change, because I believe it is the truth. What will change is that I will not single out the Jewish people politically, culturally, or in any like manner. If I should ever again write about the Jews, it will only be on a purely theological basis, and it will be said in the most non-offending manner I can muster. But to be completely open, with what has been revealed to me (and I cannot say what it is), I simply have no interest in dealing with the Jews ever again. Let's just say, there are much bigger fish to fry and leave it at that.[66]

What prompted him to write this? The answer: his motion picture. From 2002 to 2013, Sungenis wrote scores of articles and letters against Jews and their supposed conspiracies, chiefly for his website but also for *Culture Wars* magazine and other venues. These years paralleled his growing interest in geocentrism, even though the topics have no obvious connection other than eccentricity.

By 2011 Sungenis had resolved to produce a film

66. E-mail to Michael Forrest (Nov. 22, 2013).

promoting geocentrism. More precisely, he had decided to produce at least two films. The first would set the ground-work. Its purpose would be to cast into doubt modern theories of cosmology, the goal being to leave viewers with the idea that modern science is incapable of determining whether the Earth orbits the Sun or the Sun orbits the Earth. The subsequent film (or films) would promote geocentrism overtly, partly through scientific arguments and partly through theological and scriptural arguments.

As his first film, *The Principle*, began to take shape, Sungenis realized that he faced a public relations dilemma. For the film to be a success financially and in terms of influence, it would have to be screened in theaters throughout the country, and it would have to get good press. Reviews of the film would need to focus on the film, not on its producers—particularly not on him. This is where Sungenis had a problem.

For a decade he had been writing against Jews, blaming them for America's social, political, and cultural ills. He accused them of all manner of fanciful conspiracies. One was that Jews—in particular, Israel's spy agency, the Mossad—were responsible for the destruction of the Twin Towers at the World Trade Center. To an inquirer he replied, "I spend more time on the Jews because they spend more time on us. Hindus, Jehovah's Witnesses, Christian Science [sic], and Muslims have not infiltrated our Church, our government, and our culture anywhere near what the Jews have." He acknowledged that he had written "dozens of other articles related to these issues," and he gave as his motivation that "it is the typical Jewish racist mentality that I am crying out against today."[67]

67. E-mail to Keith Wasser (Aug. 14, 2012).

Sungenis had left a long, inglorious record of hostility toward Jews. He did not consider that hostility to amount to anti-Semitism—a term to which he gives a private definition—but he realized that most Americans likely would see it that way. He knew that the success of his film, in large or small part, would rely on his working with and through people in Hollywood—people who probably would hesitate to be associated with someone who, as he did, claimed that the Hollywood motion picture industry is controlled by Jews who are out to subvert the Catholic Church.

His solution was George Orwell's memory hole. By the end of 2013 Sungenis's apologetics website was transformed utterly. Gone were all of his articles, whether on Jews or other topics. Gone were links to conspiracist websites, of which there had been many. The Catholic Apologetics International website became nothing but a storefront for Sungenis's books. At this writing twenty-one titles are listed there: seven on geocentrism (nine volumes in all), four on apologetics (written in the 1990s), six volumes of Bible studies, and four miscellaneous works, including *Bob's Dictionary of Big Words*.

(Sungenis's anti-Jewish writings had been gathered and made available to the public at SungenisAndTheJews. com, a website run by Michael Forrest, who once worked with Sungenis. The website has become a shell. As Forrest explained in November 2013:

> Robert Sungenis recently contacted me and stated that he has removed [from his website] all the material about Jewish issues that led to the creation of this website and that he will not be returning to such

material in the future. As such, I've chosen to remove the documentation formerly found at this website.

This is not to imply that Sungenis has retracted and/or apologized for the statements on Jewish issues that were formerly documented here. He has not. Instead, he has recently stated, both publicly and privately, that he believes God has given him a new vision/direction related to the issue of geocentrism.

As a result of his desire to pursue this new vision/direction, Sungenis writes, "I've publicly declared that I am no longer addressing [Jewish] issues and don't wish to discuss them with anyone and . . . will never discuss them again." However, he has said that he still personally holds to the same beliefs and considers them to be true.

Forrest is monitoring Sungenis's online and periodical writings and says that, if Sungenis begins again to write against Jews, he will flip the switch and turn his website back on.)

In his reply to David Palm's "Top Geocentrists Caught Plagiarizing," Sungenis says that *The Principle*, "of which I am the executive producer, has nothing to do with the Jewish religion or the Jewish people." This is true. He continues:

As a Catholic theologian and apologist for the last twenty years, I have critiqued the religious and political positions of Catholics, Protestants, and Jews, but that in no way makes me anti-Semitic any more than it makes me anti-Catholic. Anti-Semitism is dictionary defined as an irrational hatred of the Jewish people simply because they are Jewish.

These sentences are misleading. While it is true that on occasion Sungenis criticized Catholics and Protestants, such comments were infrequent from 2002 through 2012. Most of his writings about Protestantism, particularly the Fundamentalist strain, appeared in the 1990s and focused on biblical interpretation and involved no conspiracy theories. Most of his criticisms about Catholics have been leveled at Catholics who criticized his anti-Jewish commentary. For more than a decade, nearly all of Sungenis's writing focused on two topics. His book-length work chiefly was about geocentrism, while his ephemeral writings—at his website, at other websites, and in *Culture Wars*—chiefly were in opposition to Jews and in support of conspiracy theories, with many of the conspiracies allegedly masterminded by Jews.

His words above are misleading also in the definition they give for anti-Semitism. Sungenis repeatedly has said he ought not to be labeled an anti-Semite because he does not fit what he claims to be the definition: he does not harbor "an irrational hatred of the Jewish people simply because they are Jewish":

First, let me begin, for the umpteenth time, by denying the charges of anti-Semitism. Anti-Semitism is a hatred for the Jewish race. There isn't a bone in my body that feels that way about Jewish people. . . . I have explained over and over to them that I love

Jewish people. I grew up with them, went to school
with them, played and worked with them, all my life.

Sungenis gave a similar definition of anti-Semitism in the
July/August 2009 issue of *Culture Wars*: "As we have always
known, the definition of anti-Semitism is not criticism of
Jewish beliefs and actions but hatred for the Jewish race
regardless of what they believe or do."[68]

Actually, it is not true that "we have always known"
this, because it is not accurate. Sungenis is right to say that
"criticism of Jewish beliefs and actions" does not constitute anti-Semitism, just as criticism of Catholic beliefs and
actions does not constitute anti-Catholicism and criticism of
American beliefs and actions does not constitute anti-Amer-
icanism, but it is possible that such criticism could be part
of an anti-Semitic, anti-Catholic, or anti-American posture.
Where Sungenis errs is in claiming that anti-Semitism is
limited to "hatred for the Jewish race." Such hatred certainly
amounts to anti-Semitism, but it is not a *sine qua non*.

The *Merriam-Webster Dictionary* I have at home
defines anti-Semitism as "hostility toward or discrimination against Jews as a religious, ethnic, or racial group."
Note that this definition includes no reference to hatred,
though hatred certainly qualifies as hostility. Anti-Semitism
is broader than hatred of Jews, which actually is a rare sen-
timent. Usually anti-Semitism manifests itself as a lower
level of hostility.

Both Sungenis and I have written books refuting

68. Robert Sungenis, "Is the SSPX Anti-Semitic?", *Culture Wars* (Jul.-Aug.
2009), 30.

anti-Catholic attacks by certain Fundamentalists. These attacks have been against Catholic beliefs and history and sometimes against Catholics themselves. Over the last thirty-five years I have dealt with hundreds of anti-Catholic Fundamentalists, but I cannot point to more than a handful who actually hated Catholics.

Hatred for Catholics individually or as a group seemed to be no part of the attitude of most of these opponents of Catholicism, yet the descriptor "anti-Catholic" was appropriate because they opposed the Catholic faith and sought to instill in others their own "anti-Romanist" prejudices. It was not just that they disagreed with Catholic teachings. They were hostile to them and, often, to Catholics, but their hostility did not rise to the level of hatred.

On January 21, 2014, during the course of an online discussion, I offered a definition of anti-Semite taken from *Webster's Seventh New Collegiate Dictionary*: "one who is hostile to or discriminates against Jews." An hour later Sungenis replied with what he thought was proof to the contrary—no fewer than eight definitions culled from other sources. He prefaced his list with this comment: "I could have guessed that your research would be limited to one that favored your view of anti-Semitism. Try these definitions out for size. Notice how wide the range in some of them, yet most agree that anti-Semitism is a hatred of the Jewish race."

The first definition was from the online version of the *Merriam-Webster Dictionary*: "hostility toward or discrimination against Jews as a religious group or 'race.'" The second definition was from the *Encyclopedia Britannica*: "hostility toward or discrimination against Jews as a religious or racial group." Then came the definition from *Webster's*

Third International Dictionary: "hostility toward Jews as a religious or racial minority group often accompanied by social, economic, and political discrimination." The fourth was from the *American Heritage Dictionary of the English Language*: "hostility toward or prejudice against Jews or Judaism; discrimination against Jews."

From *Webster's College Dictionary* came this definition: "discrimination against or prejudice or hostility toward Jews." From the website of the Anti-Defamation League: "belief or behavior hostile toward Jews just because they are Jewish." From Wikipedia: "prejudice, hatred of, or discrimination against Jews for reasons connected to their Jewish heritage." From the book *Ologies and Isms*: "an attitude or policy of hatred and hostility toward Jewish people."

Notice that of the eight definitions provided by Sungenis, the first six—including all of the standard dictionaries cited by him—make no reference at all to hatred. They define anti-Semitism as hostility toward Jews. Only the last two definitions use the word "hatred." Wikipedia uses it in the alternative: "prejudice, hatred of, *or* discrimination," meaning that hatred is not necessary for a finding of anti-Semitism. Only *Ologies and Isms* uses the conjunctive: "hatred *and* hostility."

So, of the eight definitions provided by Sungenis to demonstrate that no charge of anti-Semitism can be levied absent a showing of hatred, seven argue against his thesis and only one in favor. Yet he prefaced his list with the assertion that "most [of the definitions] agree that anti-Semitism is a hatred of the Jewish race." One out of eight is not "most."

What is of particular interest here is not so much the proper definition of anti-Semitism but the mindset displayed by Sungenis. So intent is he on avoiding the label of anti-Semite that he cobbles together evidence that contradicts his own position. Instead of citing the one dictionary that says that hatred is a required element of anti-Semitism and being satisfied with that as a rejoinder, Sungenis piles on seven other definitions, each of which works against this thesis and in favor of the thesis he opposes. It is as though he thinks the volume of words will overpower what the words actually say.

This suggests why his chief geocentric work, *Galileo Was Wrong: The Church Was Right*, is so massive. Its three fat volumes include about 1.2 million words, about fifteen times the word count of *Paradise Lost*. (What Samuel Johnson said of Milton's epic applies to *Galileo Was Wrong*: "None ever wished it longer than it is.")

It is impossible to turn the pages of Sungenis's books without thinking that the overall thesis would have been presented more persuasively if it had been presented more concisely. Innumerable assertions are "proved" not through a single apt argument but through a cascade of repetitive, overlapping, and sometimes contradictory quotations that seem not to have been compared with one another. Worse, many quotations are taken out of context or are truncated in such a way as to leave an incorrect impression.

Page after page, proof is sought not by the citing of one authoritative source or through a single quotation but through the sheer number of words. The effect brings to mind Hamlet's reply to Polonius: "What do you read, my lord?" "Words, words, words."

CATALOGUE OF CONSPIRACIES

ROBERT SUNGENIS SUBSCRIBES not to one or two conspiracy theories but to many. He believes the sinking of the Titanic was no accident but was a long-range blueprint for 9/11.[69] He asks, "Has modern science found irrefutable evidence that dinosaurs co-existed with humans? Yes, the evidence has been found but it is being systematically suppressed."[70] Regarding putting men on the Moon, he says, "Any intelligent person who has studied the issue is going to have doubts as to whether the United States had the capability to put a man on the Moon in 1969."[71] He says that the Moon landings were faked, secretly filmed in a studio by director Stanley Kubrick, whom he notes was a Jew.[72] And then there was Pope John Paul I, whom Sungenis

69. "The Blueprint for 9/11," n.d., formerly at catholicintl.com. In 2013 Sungenis removed this and other articles he had written from his website, catholicintl.com, in an effort to disguise his literary track record.
70. Galileowaswrong.com/wp-content/uploads/2013/06/Pope-Bendicts-Recent-Statement-on-Evolution.pdf.
71. Web.archive.org/web/20120121105745/http://www.galileowaswrong.com/galileowaswrong/features/4.pdf.
72. "Stanley Kubrick Hired to Fake Apollo Moon Landings" (Jan. 16, 2012), formerly at catholicintl.com.

thinks was murdered.[73] He also thinks that recent popes have conspired to hide the truth about the Fatima secrets.[74]

Most of the conspiracies to which Sungenis has subscribed have involved Jews: Jews were behind the assassination of John F. Kennedy.[75] Jews sent Monica Lewinsky to compromise Bill Clinton because he was insufficiently friendly to Israel.[76] Most importantly, Jews are trying to control the Catholic Church precisely because they reject the Messiah and so reject the Church he founded.

"You still don't get it, do you?" wrote Sungenis to a correspondent. "Everyone is entitled to his opinion. So cut the conspiracy jibe [sic]. Here's the point. Heliocentrism cannot be proved; it can only be postulated as an alternative scientific theory. Since it cannot be proved, then those of us who wish to take Scripture at face value when it says the Sun moves and the Earth stays still have every right to do so without being branded as a [sic] conspiracy monger."[77]

This was in reply to the correspondent's comment: "Talk about providing evidence! What a joke! Every piece of evidence provided is written off as part of the heliocentric conspiracy." But such a conspiracy is exactly what Sungenis believes in. He titled his novel *The Copernican Conspiracy*. The book is based on the belief that mainstream scientists know that geocentrism is true, that their experiments intended to disprove it instead have substantiated it, and that these scientists have suppressed the facts

73. Geocentrismdebunked.org/piling-on-or-holding-back-record.
74. Ibid.
75. En.wikipedia.org/wiki/Robert_Sungenis.
76. Sungenisandthejews.blogspot.com/2008_08_01_archive.html.
77. Robert Sungenis, "The Fallen Star: Relativity Meets the Absolute" (2002), formerly at catholicintl.com/epologetics/geofallenstar.html.

and have promoted the false theory of heliocentrism the better to protect their own livelihoods and reputations. It is a conspiracy on a grand scale, involving countless thousands of scientists over four centuries. So, yes, Sungenis is free "to take Scripture at face value when it says the Sun moves and the Earth stays still," but he is not free to deny himself the label of "conspiracy monger" when in fact he mongers conspiracies.

In July 2012 Sungenis was interviewed by Mark Dankof, who is connected with conspiracist and anti-Semitic publications and websites. The chief argument that Sungenis made on Dankof's podcast is that Israel was responsible for the 9/11 terrorist attacks and was able to pull off the attacks because Jews control America and have controlled it for more than a century:

> The Woodrow Wilson cabinet was completely run by Jews. The same was true with Franklin Delano Roosevelt. He had 93 Jews working for him. His cabinet was mostly Jewish. It was run by [Bernard] Baruch and [Louis] Brandeis, who were both Jews. . . . 9/11 basically is the capstone of it. It basically is the symbol that the Israelis are giving us that "We are in complete control over your country and that there's nothing you can do about it."

This is nonsense. Most of the twenty people who served in Wilson's cabinet were Protestants, perhaps the best known being William Jennings Bryan. Not one in his cabinet had a name that might suggest Jewish ancestry, except for a fellow surnamed Glass, but his mother's maiden name was Christian. The names of all the other members of Wilson's cabinet seem to be English in origin. Of the twenty-four people who served in Roosevelt's cabinets, only one, Henry Morgenthau, was Jewish.

It is true that Bernard Baruch was a Jew. He was a financier who served as an advisor to Roosevelt, but he never served in the cabinet. Louis Brandeis, also a Jew, was an associate justice of the Supreme Court. He never served in the cabinet either. It is unclear how Roosevelt's cabinet could have been run by two men who did not even attend cabinet meetings.

Sungenis is convinced that Jews desire to run the world at the expense of Christians. In *Culture Wars* he asked, "Are we to pretend that Jews have never tried to dominate the world and defeat Christianity by accumulating money and political power to promote their anti-Christian and humanistic worldview?"[78] Elsewhere he said, "As for Germany's relationship with the Jews, well, the Germans treated the Jews very nicely when the Jews were excised out of Russia and migrated to Germany. Then the Jews turned on the Germans because they got a better deal from someone else."[79]

Such comments induced some people to dispute with Sungenis. He had a long e-mail exchange with inquirer Keith Wasser, who declined to subscribe to Sungenis's conspiratorialist views. After many e-mails were sent from one to the other, Sungenis said, "We've finally gotten to the core

78. Robert Sungenis, "Is the SSPX Anti-Semitic?", *Culture Wars* (Jul.-Aug. 2009), 33.
79. Goodreads.com/author_blog_posts/781638-so-the-other-day.

of your argument. Unfortunately, it is the typical Jewish racist mentality that I am crying out against today. This is precisely why you can't see the truth about the Jews and Israel, and it is precisely why you don't like me pointing the finger at them." Later, still more frustrated with his correspondent, Sungenis wrote this to him: "I received your next e-mail. Before I answer your question, I would like to know one thing: Are you Jewish? Wasser has a Jewish origin, from the sources I have investigated."[80] ("Wasser" is German for "water.") This was not an isolated incident. Sungenis has accused other critics of concealing their Jewish lineage.

The remark to Wasser was made in 2012, when *The Principle* was in production and when Sungenis was beginning to see that he would need to take down from his website his multitudinous anti-Jewish comments. Someone he trusted must have said to him that such comments—not just those in public but also in private correspondence—would not be good for the film's public relations. The same could be said of his other conspiracy theories.

Among Sungenis's more outlandish assertions is that crop circles are made by lasers or plasma beams shot from satellites controlled by NASA. When called on the claim, he responded: "The fact is, crop circles do exist and we must give a rational explanation for them rather than pretend they do not exist. The alternative explanation (e.g., that they were created by aliens) is much more bizarre and dangerous than believing that someone on Earth made them."

He writes as though there is but one alternative

---

80. Robert Sungenis e-mail to Keith Wasser (Aug. 14, 2012).

explanation: extraterrestrials. He overlooks the most widely accepted explanation: that crop circles are made by pranksters using planks of wood and rope. (In England, two men claimed to have made hundreds of crop circles—and demonstrated to news media how they did so.) Close examination of a crop circle shows that the crop, such as wheat or corn, has been broken within inches of the ground and has been arrayed in a circular pattern, just as if a long plank, fixed at one end, were dragged in a wide circle. Crop circles invariably appear near roads and populated areas; they always provide easy access for the curious.

Why would NASA want to make crop circles? Sungenis has an explanation:

I think this whole thing is cooked up by NASA and the powers-that-be in order to control people. I also think crop circles can be made from space with lasers or plasma projectors. All NASA would have to do is put a digital pattern in a laser/plasma projector aboard a satellite and then shoot it down to earth, and presto, you have a crop circle. It gets everybody talking about UFOs. But really, all they are doing is getting our minds off the Bible and Christ by making it look like neither are [sic] true.[81]

This seems like an inefficient method to get people's minds off religion. Wouldn't it be cheaper and more effective for NASA to print and distribute, in stealth, millions

81. Web.archive.org/web/20120118153827/http://bellarmineforum.xanga.com/702646935/question-139---what-do-you-think-of-ufos.

of anti-religion leaflets? It also seems like an impossible method of creating crop circles. How could a laser or plasma beam, sent from hundreds of miles above the surface of the Earth, reach the base of the stalks and cause the plants to bend precisely there while not frying the upper parts of the plants? How could a beam cause all the plants to bend in a consistently clockwise or counter-clockwise direction? Why wouldn't the plants, their bases severed, fall helter skelter, making no pattern?

Besides, what possibly could be NASA's motive for doing such things? Sungenis knows: "NASA has every incentive in the world to promote UFOs, simply because they want to implant in our heads the idea that there is life on other planets. In that way the government will continue to give billions of dollars to a program that the government has thought more than once of scrapping."[82] Yet there is a deeper reason still for NASA to "promote UFOs":

The very reason popular science wants you to think the Earth is somewhere far away in the remote recesses of space is so they can promote the idea that we are not special. If we are not special, then surely there must be other beings in the universe, too. All of this, of course, ties in with evolution, because evolution explains how we got here. It's all a big, big lie cooked up by the Devil.[83]

Sungenis continues his excoriation of NASA in

82. Ibid.
83. Ibid.

"Catholic Traditionalist Struggles with Geocentrism,"[84] a 47,000-word reply to a 13,000-word blog post. Sungenis's piece was uploaded to GalileoWasWrong.com on November 11, 2014. He said the young man to whom he was replying "doesn't realize that the quest to find alien life on other planets is part-and-parcel with the Copernican Principle to keep Earth out of the center of the universe and be regarded as nothing special," and this "shows just how well NASA's program has worked to convince Christians that NASA has the true story and the Church is full of ignoramuses."

Sungenis indicts NASA—and apparently nearly all of its employees—for plotting against the Church. It is a plot the existence of which he infers but not one for which he can adduce the slightest evidence from the organization itself. Are there no telltale e-mails from Christian whistle-blowers within NASA? Has no one leaked an internal memo? Has no one working there, over the more than half century since NASA was established, had a change of heart and confessed to the organization's true motives when requesting funding for such projects as the Pioneer 10 and 11 spacecraft, which carried engraved plaques meant to inform extraterrestrials where the satellites came from?

"If we are not special": here is the reason that the slogan for Sungenis's film *The Principle* is "Are you significant?" If extraterrestrials exist, and if they visit Earth in their UFOs, the answer would be "No, we are not special." Man would not be uniquely significant—and perhaps not significant at all—if on other planets or in other galaxies there were

84. Robertsungenis.com/gww/features/Catholic%20Traditionalist%20 Struggles%20with%20Geocentrism.pdf.

creatures more highly developed, intellectually or scientifically, than we are. But one need not worry about any of that, says Sungenis. "There are no aliens from other planets. The universe was made for this Earth only. That's not hard to understand once you accept that the Earth is in the center." It is from the physical centrality of the Earth that flows our confidence that man is supreme in the visible, created world and that nothing else in the universe can impinge on his status.

All this speculation about extraterrestrials, UFOs, crop circles, and NASA's machinations comes down to preserving the physical centrality of the Earth, without which, according to Sungenis, man would be unimportant. In this he errs. He assumes that importance—to whom: man or God?—is a matter of geography, when, in the Christian understanding of creation, importance is a matter of God's love for man, whom he created in his own image and likeness.

This catalogue of odd beliefs is not given here to embarrass Sungenis but to suggest that a man who is so unreliable in his judgments and so suspicious in attributing motives cannot be relied on when explaining matters of science, history, or theology.

It is one thing for a man to have a noted eccentricity, either in personal comportment or in belief: the fellow who invariably carries a walking stick though he has no difficulty in walking, the man who knows the location of the Lost Dutchman's Mine but isn't telling, the inventor who really has discovered the key to perpetual motion. It is something else to subscribe to a jumble of conspiracies arising from suspicion or prejudice. When many of these conspiracies involve matters of science or engineering, and when those matters are misunderstood (such as what lasers can and cannot do), the conspiracist's cognitive and deductive powers properly are called into question.

9/11 AND THE GEOCENTRIC MIND

I N THE LAST thirteen years, probably nothing has resulted in more column inches of Internet disputation than have the events of September 11, 2001. Robert Sungenis has not refrained from offering his own opinions. They put him squarely in the camp of the so-called "9/11 truthers." To the usual conspiracy theories he has added elements of an obsession with Jews. His most extensive writings about the destruction of the towers at the World Trade Center came in a lengthy review of Christopher Bollyn's *Solving 9/11: The Deception That Changed the World.*[85]

The review was so long that it was split between two consecutive issues of E. Michael Jones's *Culture Wars* magazine. Ten pages of the review appeared in the July/August 2012 issue, and a further fifteen pages appeared in the September 2012 issue. The title was "Inside Job." Sungenis's review is worth examining, even though it does not concern geocentrism, because it shows us something about his thought processes. He uses those same thought processes when evaluating scientific evidence.

85. Christopher Bollyn, *Solving 9/11: The Deception That Changed the World* (Christopher Bollyn, 2012).

Bollyn's position, which Sungenis accepts almost entirely, is that the Twin Towers were brought down by explosives that were planted and detonated by Israel's intelligence service, the Mossad. Sungenis rejects the idea that the culprits were Muslim terrorists. Such people were not sophisticated enough to pull off such a large-scale operation, he thinks. But the Israelis were. He makes his most sustained, and perhaps his most moving, argument in a long series of rhetorical questions:

Who had the power to pull it off without a hitch? Who had the most pressing motive? Who had the money to pay for it all? Who had control of the military and NORAD to force it [sic] to stand down while four passenger planes went to their designated targets? Who knew the intricacies of U.S. commercial flights to get around the FAA air traffic control? Who had the power, if needed, to operate planes by remote control?

Who had control of the courts to make sure that no wrongful death claims went to trial? Who had control of disposal so that all the steel girders of the Twin Towers were shipped to China before they could be chemically analyzed? Who had the power to corral the NYPD and NYFD? Who had the power to allow dozens of suspects to escape to foreign countries? Who had the technology to bring down steel-girded buildings built to withstand forces much greater than aberrant planes and nominal fires? Who had the advanced computer knowledge to coordinate the attacks? Who had state-of-the-art knowledge about explosives and detonators?

Who had control of the press and media to curtail investigations? Who is most familiar with and has political control over the city at which the attacks took place? Who has a history of unprovoked attacks or even false-flag operations? Who has a network of spies operating in the U.S. that could facilitate the attacks? Who has personnel in almost every sector of the federal government to create such a master cover up? Lastly, as any detective would ask, who benefits from the attacks?

This avalanche of questions will impress some people by its sheer weight, but when the questions are examined one by one, the rhetorical power declines. Some of the questions illustrate Sungenis's ignorance of how things work.

He asks, "Who knew the intricacies of U.S. commercial flights to get around the FAA air traffic control?" Only a non-pilot could ask such a question. Air traffic control is unable to exercise any control at all over a plane piloted by someone who ignores ATC's instructions. ATC is like a traffic cop standing in the middle of an intersection. He can wave his arms to direct traffic, but he can do nothing about a car whose driver ignores him and drives right through.

Sungenis asks, "Who had the technology to bring down steel-girded buildings built to withstand forces much greater than aberrant planes and nominal fires?" What forces could those be? Even earthquakes, which have not been known to occur in Manhattan, would not exert on the upper floors of a tall building the force provided by a crashing airliner, and no storm, not even the most violent nor'easter, would come close to such a force either. The Twin Towers were not built to withstand the impact of airliners, just as they were not built to withstand the impact of asteroids. Those possibilities were considered too remote. The towers *were* built to withstand the impact of small planes because such planes had been known to fly into skyscrapers.

A few pages further into his review Sungenis says it "is insulting to one's intelligence" to "pin 9/11 on some sandal-wearing vigilante [Osama bin Laden] hiding in caves of Afghanistan and nineteen Muslims who wouldn't know how to fly a commercial jet liner into a building if their life [sic] depended on it." We can ignore the unintentional humor of the final clause, but it is worth challenging Sungenis's assumption that the nineteen could not have been capable of maneuvering airliners. That just isn't so.

At least two of the terrorists obtained instrument ratings through a flight school at Montgomery Field airport in San Diego. I received my own Airplane Single Engine Land rating at that field, though from a different flight school. The terrorists did not need to know how to control an airliner in the take-off or landing configurations; they needed to know only how to maneuver it while it was in the air. They obtained sufficient information through their instrument-rating instruction.

The instrumentation on the panels of airliners is more complicated than that on the panels of private aircraft, but most of that instrumentation relates to things other than the steering of the planes. For that it is sufficient to have knowledge of the on-board GPS (Global Positioning System), which is not greatly different from that found in automobiles. All the terrorists would need to have done is to select the name of an airport near their target—La Guardia, for example—and then manually steer the airliner to follow the course shown on the screen. Once they had Lower Manhattan in sight, they could steer visually since the Twin Towers were easily seen at a great distance.

In fact, they didn't even need to use on-board GPS at all. It would have been enough for them to bring aboard a handheld aviation GPS. These are similar in size to GPS units used by backpackers, except, instead of having ground maps, they show airspace boundaries and the locations of airports. I have flown considerable distances using a handheld unit. It is easier to follow the course shown on an aviation GPS, whether handheld or panel mounted, than to follow a road shown on a GPS used by drivers. Roads meander, but pilots just have to fly in a straight line. Sungenis, ignorant of the basics of flying, leaves his readers thinking that something that is simple is so complicated that it could not have been done by accomplices of a "sandal-wearing vigilante."

He makes other groundless comments, most of them one-liners that he makes no attempt to substantiate and all of them easily refuted. For example, he says that "no passenger plane debris was ever found" outside the Pentagon—at least not immediately. "The plane debris was placed on the grounds and the light poles were knocked down after the 9/11 event. . . . The eyewitnesses further state that it was indeed a plane that came toward the Pentagon but one that shot a missile at the Pentagon and then the plane ascended and flew over the Pentagon." All of this is false. Much debris was found by first-responders, and many people saw an airliner coming in low and hitting the side of the Pentagon. Sungenis denies this. He even doubts that commercial airliners hit the Twin Towers at all; more likely, he says, what struck them were "remotely controlled tankers camouflaged to look like commercial jet liners."

He thinks it ominous that "many 9/11 witnesses have been mysteriously dying since 2001," but one should expect, over more than a decade, that some of the people who saw the destruction of the Twin Towers—and there were thousands of such people on the streets of New York on September 11—would go to their eternal reward. Did some of them die "mysteriously"? Possibly so, but every day people die in inexplicable circumstances. There is no proof that anyone lost his life as retribution for his being a witness to the events of September 11.

Sungenis thinks that Flight 93, which crashed in a field in Pennsylvania, "was supposed to hit Building 7" of the World Trade Center and that, once it failed to do so, the Jewish owner of that building ordered the building's destruction. But how could any plane, except a very small one, have been maneuvered to hit Building 7, which was 47 stories tall (less than half the height of the Twin Towers) and in an area of taller buildings? To strike it, a plane would have had to descend steeply, something no jetliner is capable of doing. (The planes that struck the Twin Towers were in almost level flight at the moment of impact.)

Perhaps the most bizarre allegation written by Sungenis, as part of his argument that Mossad agents arranged the destruction of the Twin Towers, is his claim that the agents were seen celebrating their triumph in three widely-separated locations: "Witnesses saw them jumping for joy in Liberty State Park after the initial impact. Later on, other witnesses saw them celebrating on a roof in Weehawken, New Jersey, and still more witnesses later saw them celebrating with high fives in a Jersey City parking lot."

How could someone write this with a straight face? Weehawken is six miles north of Liberty State Park, which lies opposite the southern tip of Manhattan. From Weehawken, it is another four miles back down to Jersey City. Are we to believe that Mossad agents moved from place to place, each time showing spontaneous expressions of jubilation, jumping for joy in one place, celebrating in another, giving high fives in a third? And could these really have been agents of the Mossad, an organization famed for stealth? It is possible to imagine Mossad agents smiling in satisfaction within the confines of a closed room, but it is ludicrous to think that experienced operatives would make public displays of triumph in three separate locations.

Sungenis believes that Larry Silverstein, owner of Building 7, was "the key to 9/11. Since Larry gave the order to bring down WTC 7, either he or someone else gave the order to bring down the Twin Towers in which 3,000 people were roasted alive, and it has nothing to do with nineteen Muslims, most of whom are still alive. This is the biggest hoax in history."

In his long book review in *Culture Wars* and in subsequent and sometimes lengthy letters to the editor, Sungenis shows himself to be susceptible to the bad thinking of others. He discounts none of Christopher Bollyn's claims; he expresses no reservations; he confides no doubts. He knows Jews were responsible for 9/11 and that Muslims were not, except perhaps as patsies. Larry Silverstein was in it for the insurance money, even though he could have profited as much simply by selling the Twin Towers to someone else—and he would not have run the risk of public exposure by investigators such as Bollyn and Sungenis. Overconfident, Silverstein, in cooperation with the Mossad, undertook an operation that backfired. He and it have been shielded from prosecution only because Jews control the legal system.

That is the mindset of Robert Sungenis. His comments about 9/11 tell us something about how he thinks, and they tell us something about his approach to evidence and the lack of evidence. His long obsession with Jews has inclined him to look for a Jewish connection whenever he is confronted with an unsavory event. Often enough he thinks he finds one, and he manages to give facts (and factoids) just the interpretation needed to implicate Jews, even though more likely and exonerating interpretations are possible.

If he has shown himself untrustworthy in evaluating what happened on September 11, 2001, is it prudent for his readers to presume he is trustworthy on other matters, particularly in subject areas—such as science—in which he has no training beyond that of most undergraduate students?

Is it prudent for his readers to trust representations and interpretations made by Sungenis's associates, people who presumably have no great reservations about his ideas or methodologies? If Sungenis often is off the mark when speaking about history or politics or science, what about his colleagues, such as Rick DeLano and Mark Wyatt? How much confidence should be reposed in them if they repose confidence in him?

FICTION TEACHING FICTION

ROBERT SUNGENIS'S CHIEF work on geocentrism is *Galileo Was Wrong: The Church Was Right*,[86] the tenth edition of which appeared in April 2014. It will not be read in a weekend even by a speed reader. The three-volume set totals exactly 2,200 pages in a format of six-by-nine inches with closely-spaced text and narrow margins. There are 1.2 million words altogether. Occasional illustrations break up some pages, but that visual relief is overwhelmed by interminable footnotes set in a small font.

Sungenis may have saturated his market early on. His trilogy languishes around the two million mark at Amazon's best sellers ranking, meaning that two million books are better sellers. No doubt his most devoted followers purchased *Galileo Was Wrong* years ago, when it appeared as two fat volumes instead of the later three, and most of them probably have seen insufficient reason to purchase the latest, expanded edition, which is available at his website for $114 or at Amazon for $94. Potential buyers who are

---

86. Robert Sungenis, *Galileo Was Wrong: The Church Was Right*, tenth ed. (State Line, PA: Catholic Apologetics International, 2014).

outside of Sungenis's core group, if they come across the set at all, likely are put off by its length, density, and dry text. (*War and Peace* is less than two-thirds as long and boasts better writing, character development, and a plot.)

Seeing that his masterwork could not be sold in volume, Sungenis brought out two variants early in 2014. The first was *Geocentrism 101*,[87] which is subtitled *An Introduction to the Science of Geocentric Cosmology*. It is "recommended for high school, college, and adult education." The back cover explains that this is "a layman's version" of the full set, which has been "condensed into this 230-page book for your personal learning and enjoyment." (The book actually has 244 pages.) Producing *Geocentrism 101* makes some sense. It is harder to see the sense in Sungenis's other attempt at popularization, a novel called *The Copernican Conspiracy*.[88]

The back cover of the novel invites the reader to "come on a life-changing journey with college sophomores Joshua Gemelli and his life-long sweetheart, Mary McGinnis, along with their newfound friend, Suresh Verma, as they seek for academic freedom at St. Robert Bellarmine College" (not to be confused with real-life Bellarmine University, located in Louisville, Kentucky). Like everyone else, the three students were brought up thinking heliocentrism to be true, but, "to their utter amazement," they "discover that the very physics courses they are studying at the college, if given the most obvious and fitting scientific interpretation, show the Earth is motionless in the center of the universe."

87. Robert Sungenis, *Geocentrism 101* (State Line, PA: Catholic Apologetic International, 2014).
88. Robert Sungenis, *The Copernican Conspiracy* (State Line, PA: Catholic Apologetics International, 2014).

How do they make this discovery? It begins on the first page of the novel. Josh and Mary are leaving the classroom. Professor Samuel Richenstein has just explained (or explained away) the 1887 Michelson-Morley experiment. Josh senses that something is wrong. Richenstein too easily dismissed the idea that the experiment may have demonstrated that the Earth is at the center of things and is itself motionless.

That's all the prompting the students need to head for the library, where they quickly discover that the truth of geocentrism "is clearly stated in the scientific literature but has been ruthlessly covered up as part of a worldwide conspiracy to keep the truth from coming out." (If there is a worldwide conspiracy to hide the truth, why is the truth so easily found by a pair of college students? Why didn't the conspirators removed the truth from the scientific literature—or at least from the college library? The conspirators must be among history's most incompetent.)

Josh and Mary realize that "they have just what the world needs—a whole new and better perspective with which to view themselves and the God who created them. They are full of vim and vigor, anxious to begin their new lives."

"Vim and vigor"? The novel is full of such hackneyed phrasing; on the third page, for instance, the students are called "a dynamic duo." Worse, the novel is devoid of character development. The reader learns almost nothing about the protagonists—Josh and Mary struggle to reach two dimensions—and ends up caring little for them as people. They are stick figures whose temporary disappointments ("Allowed to think by herself for a moment, suddenly Mary's castle in the sky came crashing down") hardly can move the reader's heart.

Throughout the book Sungenis violates the first principle of fiction writing: show, don't tell. Don't say, "John looked angry." Say, "John's lip curled." Don't say, "Sally felt relieved." Say, "Sally let out a long, slow breath." There is almost none of that in *The Copernican Conspiracy*. The characters are described for the reader; the reader doesn't see the characters from the inside, and thus the whole novel reads flat.

The story's villain comes off little better than its heroes. Samuel Richenstein has a surname as obviously Jewish as Josh's and Mary's surnames are Catholic. (The choice of "Richenstein" highlights the author's long-time obsessions with Jews and conspiracies supposedly led by Jews. Given Sungenis's recent attempts to distance himself from a decade of anti-Jewish writing, one might suppose he would have chosen for his heroes' nemesis any name other than a Jewish one.) Only in one passage do we catch a glimpse of Richenstein as a human being:

> The professor's words were a consequence of the indelible mark that had been placed on him thirty years prior when he was faced with a similar situation as a student challenging the scientific dogma of the day—a situation in which he, at that time, had decided to become a member of the establishment and to extinguish the childlike inquisitiveness that got him into science in the first place. Science had now become a job, not an adventure. To maintain one's job meant maintaining the status quo, and never looking back. As far as he was

concerned, it took decades to put his sheepskin on the wall and secure a six-figure salary.

Josh is a faster learner than Richenstein. By page five he has concluded that there has been "a deliberate attempt by the professor to sweep a perfectly good solution to the [Michelson-Morley] experiment under the rug so that modern science doesn't have to face the truth that it has been wrong for over five hundred years!" Josh—named after the biblical character for whom the Sun stopped (Joshua 10:13)—is a transparent stand-in for Sungenis himself, who described his own mission this way when replying to an Internet inquirer:

> I am pursuing one of the greatest projects the Lord has ever given me. It is to tell the world that the Catholic Church was right when it condemned Galileo, and thus no Catholic has to hang his head in shame. *That* is why I am pursuing the Galileo issue . . . because the right view of it can change the world and restore the Church to her rightful place of honor. I'm going for the whole enchilada. If *Galileo Was Wrong* and the Church was right, you can imagine what an impact that will have on our whole view of ourselves and the modern age. That is what I call the ultimate "apologetic," and that is what I will be pursuing the rest of the time the Lord gives me.[89]

The Copernican Conspiracy is not so much a novel as a long lecture. In this it has similarities to *Atlas Shrugged*, but

89. E-mail to Wes Grant, n.d.

without Ayn Rand's literary pretensions. Its text is punctuated by more than seventy illustrations. Most of them are images of covers of books discovered in the library by Josh, Mary, and their India-born friend Suresh, or they are images of pages from those books or from holographic letters written by key figures in nineteenth- and twentieth-century cosmology. When a page from a book is reproduced, Sungenis underlines or boxes the text to which he wishes to draw attention. The three students pore over the items they have discovered, and soon the novel's male lead has a revelation.

> Suddenly, it seemed Josh was experiencing one of those moments when one's whole life flashes before his eyes. This was the moment of truth. Would he answer "no" [to Richenstein's "Do you believe the Earth isn't moving, Mr. Gemelli?"] and just be one of the crowd, not cause any disturbance, and keep his scholarship secure? Or would he answer "yes" and remain true to the convictions he cultivated after much research and especially after he gave that impassioned speech in the cafeteria about changing the world?

"I am not ashamed to say 'yes' to that question, professor," answers Josh. Sungenis reports, "There it was! The words came out, although they were not accompanied by the near-death experience of which Josh had been fearful just a few seconds prior." The student's "response was so simple and so short, but Josh knew instinctively that nothing would ever be the same. He was about to embark on a road of no return, but a road leading him to heights that would, indeed, be a first step in fulfilling his dream of changing the world. . . . He knew one thing for sure: he felt like a new man, even though the burdens of the world seemed to be placed on his shoulders at that very moment."

These sentences are less about Josh than about Sungenis himself. Years ago he was the one who gave a public "yes" to geocentrism. The "words came out," and nothing untoward happened, but he "knew instinctively that nothing would ever be the same." Eventually his focus on geocentrism necessitated his laying aside his extensive anti-Jewish writings; as important to him as they were, they proved to be a burden, drawing people's attention to his private fixations and away from "his dream of changing the world"—or at least of changing how people view the world's status in the cosmos.

Sungenis has found his quixotic work invigorating. He feels "like a new man, even though the burdens of the world [seem] to be placed on his shoulders." With Josh he can say, "We now know what we are up against. It appears to me that we've hit upon the best-kept secret of our modern culture."

At the end of the sixth chapter Sungenis, speaking through Josh, says a prayer:

> Thank you so much for the opportunity to take on this challenge to present your creation to a world of skeptics and unbelievers. Thank you for choosing me to reveal to a dying world that this Earth is special and put in a very special place in the universe

because you loved it and them so much. I pray that you will give me all the knowledge and wisdom I need to make a convincing case. And I ask that you open the minds of my hearers that they may face the evidence with courage and honesty.

This is the kind of prayer that Sungenis might pray when he contemplates the burden that became his following his discovery of the truth of geocentrism. It is rejection of this truth that has caused the world's current ills, yet it is a truth that is accessible to all, if only the conspiracy to hide it can be undone. (Mary tells Josh, "There is a real conspiracy here, I'm convinced.")

Later on, as the two, along with their friend Suresh, uncover further evidence of conspiracy, they come to see that the key lies in the false interpretation given to the Michelson-Morley experiment, which really showed that aether exists. As Josh explains to a confused Suresh, "Metaphysically speaking, there must be something between Earth and the Moon or the Earth and the Sun. We cannot say there is nothing between them, since if that was the case the Earth and the Moon would be touching each other."

Here it is Sungenis who falls into a metaphysical—really, a physical—trap. He thinks that empty space can't exist because its emptiness would provide no medium for the transmission of light and no way to keep objects apart. Thus aether must exist. But what is aether? Like Gerardus Bouw, Sungenis believes that it is an immensely dense substance that happens to be invisible to our senses and that fills all space not occupied by objects such as planets and stars. Much as drops of water, lying side by side in the sea, transmit from one to the next the undulating motion of the waves, so the aether, which is made of fantastically small particles, transmits the undulating motion of waves of light.

Sungenis believes empty gaps between celestial bodies can't work: "if that was the case the Earth and the Moon would be touching each other." But what about gaps between the particles that make up the aether? What is between the particles, other than empty space, albeit very small empty space? His notion of aether doesn't solve a problem he thinks he sees.

Nor does it solve a different problem, one that arises only for geocentrists. They say the star field revolves around the Earth each twenty-four hours. How can it do so without the stars moving fantastically faster than the speed of light? Sungenis thinks the aether makes this possible. Aether is so dense that it "could revolve around the Earth trillions of times faster than twenty-four hours and still remain stable"[90]—although he maintains that, in reality, the aether revolves exactly once in twenty-four hours and carries the stars, Sun, and planets with it, thus producing night and day.

It is the distant portions of the aether, not the distant stars themselves, that move faster than light. The stars are carried along by the aether except in their local movements through it. Those local movements, which are made with respect to the aether, are restricted to the speed of light. The aether itself is exempt from light's speed limit. When measuring a distant object's speed, that measurement is to be taken not in reference to the Earth but in reference to the local aether. In this way one can say that the stars, though

90. Robert Sungenis, letter to the editor, *Culture Wars* (May 2013).

circling the Earth each twenty-four hours, do not exceed the speed of light.

Sungenis is forced to make an argument along these lines—to say that something other than the stars does exceed the speed of light—because to do otherwise is to say that the stars themselves exceed that speed. His argument hinges on the idea that the aether differs so greatly in its substance from things that science can detect and measure that it is not bound by laws that bind other kinds of matter. If aether does not exist, and if it has to be the stars themselves that move around the Earth in twenty-four hours, the geocentrist would have to deny the well-established limit that is the speed of light—not only for the stars but for any body more distant from the Earth than is Saturn.

Consider the case of the closest star to us other than the Sun, Alpha Centauri. It is 4.37 light years from Earth. That is about 26.2 trillion miles. If Alpha Centauri circles the Earth each day (Sungenis says the stars revolve around the Earth but not quite in perfect circles), its path would be 161.3 trillion miles long. To cover that distance in twenty-four hours would require a speed of about 1.86 billion miles per second, yet, as even most high school students learn, the speed of light is 186,000 miles per second. This means Alpha Centauri would need to travel at ten thousand times the speed of light. This is not possible for any object detectable by scientific instruments. The solution for Sungenis and other geocentrists is to posit an imaginary substance, the aether.

Sungenis falls into other traps of his own making as he has his novel's protagonists discover further incriminating evidence. Josh stumbles "across the story about a professor of physics who was teaching at Catholic University of

America just on the other side of town. He was speaking before the American Physical Society[91] and dared to say that Einstein's Special Relativity was wrong." He lost his job as a consequence. There follows a long quotation that is footnoted to a February 10, 2014, blog post written by Charles W. (Bill) Lucas:

> I was in the faculty of the Catholic University [of America], Washington, D.C. I am not a Catholic. I gave a talk before the American Physical Society on the validity of the five principal assumptions of Special Relativity. Next day the president of CUA received a letter from the Department of Energy and the National Science Foundation demanding that I be immediately fired from my tenured position. . . .

> [T]he chairman of the physics department showed me the letter and told me I was formally fired because they needed to protect the other twenty-seven positions at their school. . . . And that's the way it's done in America. You may not know about it, but it is done and it is done regularly, and I have met hundreds of other scientists who have had exactly the same experience.[92]

Maybe not. Sungenis came upon this quotation from Lucas and accepted it without further inquiry. He should have

91. The American Physical Society, founded in 1899, is reputed to be the second-largest society of physicists in the world and the largest headquartered in the U.S.
92. *The Copernican Conspiracy*, 206–207, citing ivorcatt.co.uk/x42c.htm.

wondered whether a tenured professor could be dismissed overnight without any due process (the answer, at least at CUA, is no) and why Lucas has taken no action against the university for allegedly violating his tenure rights.

Just who is Bill Lucas anyway? At the website of Commonsense Science, a creationist organization that he is part of, his résumé says that he received a Ph.D. in physics from the College of William and Mary in 1972, and then he "performed post-graduate research on pions [subatomic particles] at Catholic University in Washington, D.C."[93] There is no mention of his being on the faculty at CUA— but there used to be such a mention ("Professor of physics at Catholic and American Universities in Washington, D.C.") until Lucas was challenged on it by an anonymous blogger who ran a website called "Defending Science, Scientists, and Non-Scientists."

The blogger said he contacted CUA and learned that Lucas never had been on the faculty. "When I confronted him over it, he said that what is on his résumé is not his responsibility and he'll get it corrected."[94] Eventually the reference to being on CUA's faculty was removed. (As the quotation given by Sungenis shows, Lucas again is claiming to have been on the CUA faculty.)

The unnamed blogger reported on a presentation Lucas gave. Lucas rejects Einstein's theories because he thinks they restrict God's powers. "He said that because of relativity God could not be omniscient. And he said that quantum mechanics does not allow for a definition of sin." Another

---

93. Commonsensescience.org/survey/popups/charles_lucas.html.
94. Silkworm.wordpress.com/2006/06/22/dr-bill-lucas-update-1-week-after-being-discredited.

website says that Lucas believes that Earth's gravity has been declining for centuries. A consequence, as the writer at that website paraphrased it, is that "people in biblical times lived so long because the intense gravity of the Earth kept more oxygen close to the surface."[95] Today's gravity supposedly is only one-ninth as strong as it was then, which explains why today's maximum longevity is around 100 years compared to the 900 years of mankind's earliest days.

So Sungenis, through his hero Josh, writes about a supposed smack-down of a geocentrist professor, accepting the man's characterization of himself, but the professor has made questionable claims about his status, and he has promoted some ideas that qualify as bizarre. (Lucas thinks Earth's gravity has declined to one-ninth of what it once was because Earth's mass has declined by the same proportion. If gravity had been nine times what it is today, no one could have survived. A 200-pound man would have felt as though he weighed 1,800 pounds. He would have suffocated.) Sungenis's alacrity to reprint anything that seems to bolster his claim of a conspiracy to keep the truth of geocentrism secret can be paralleled to his earlier alacrity to reprint material that alleged conspiracies led by Jews. The focus may have changed, but the sloppy mindset persists.

The Copernican Conspiracy ends with a promise for two sequels, *Journey to the Center of the Universe* and *The Galileo Gambit*. They may never appear, since the first book in the series is languishing around the one million mark in the Amazon rankings. That low ranking is not helped

95. Silkworm.wordpress.com/2006/06/16/the-return-of-lucas-recap-corr-june-15-2006.

by the book's back-cover blurbs, which were written by Sungenis. He promises the prospective reader that "you will laugh; you will cry; and you will even learn a little physics."

The last probably is true, though some of the physics is wrong, as are many of the inferences drawn from it. The other two promises—"you will laugh; you will cry"—likely are true too, but not in the way Sungenis expects. He thinks readers will laugh and cry in sympathy with his heroes, but the laughter and tears more likely will result from trying to make it through a dull book that is less a novel than a screed.

The other short-form book Sungenis has produced is *Geocentrism 101*. It is a simplification of the extended arguments given in *Galileo Was Wrong*. Like Sungenis's other books on geocentrism, it is self-published and looks that way. The typography is non-standard (the table of contents, running heads, and final chapter are set in a font quite different from the rest of the text), word spacing is irregular (there are many large gaps) because automatic hyphenation was turned off, block quotations sometimes are set within quotation marks and sometimes not, margins (particularly at the top) are too small, and there is no evidence that an attempt was made to follow a standard style guide, such as the *The Chicago Manual of Style*, *The Associated Press Stylebook*, or Kate Turabian's *Manual for Writers*.

The book has three hundred black-and-white photographs and illustrations, most of them indistinct. Many are computer screen captures, while others are much-reduced two-page spreads from books being referenced. The images are arranged haphazardly—sometimes flush left, sometimes flush right, and sometimes centered. Text flows around them awkwardly. Sungenis could have profited from paying a professional to lay out the book for him. It would be difficult for a prospective buyer who thumbs through *Geocentrism 101* to take it seriously because it looks like what it is, an amateur production.

The book is "recommended for high school, college, and adult," but its production values are so low that it is unlikely that many schools will order it, even if they otherwise were willing to stock a book on "fringe science," the term used by Jeffrey C. Foy, co-producer of Sungenis's film *The Principle*, to describe the genre in which the film and related publications fall. Sungenis claims to have spent a million dollars on his film but could not have spent one percent of one percent of that on designing his books, which, despite their many flaws, probably have a better chance of finding an audience than does the film.

THOSE PESKY SATELLITES

THE PROBLEM (for geocentrists) of geostationary satellites was addressed briefly when discussing Gerardus Bouw's *Geocentricity*. In that book, his major work, Bouw devotes only one page to the topic, and even then he gives not his own words but the awkwardly written words of an anonymous writer. Bouw's Catholic counterpart, Robert Sungenis, also gives just one page to the topic in *his* major work, *Galileo Was Wrong: The Church Was Right*. He has written more about it in online exchanges.

A sympathetic questioner wrote to Sungenis: "I have been thinking about the Earth being stationary and not rotating. . . . I have been unable to figure out how that system would explain communications satellites that are 'fixed' over one city on the equator of the Earth. . . . I understand how a stationary, non-rotating Earth would have a satellite launched from her that goes round and round, always changing which city it is over, but how do you get a fixed one . . . without a rotating Earth?"

Sungenis begins his long reply[96] by saying, "Thanks for

---

96. Robert Sungenis, "The Fallen Star: Relativity Meets the Absolute" (2002), formerly at catholicintl.com/epologetics/geofallensar.html.

the great question. This one really is an eye-opener." It turns out he is right—but not in the way he suspects. His answer is an eye-opener because it shows his confusion on multiple points. Of all the challenges to their theory, that of the geostationary satellites may be the most troublesome for geocentrists. The topic is worth considering at length because it demonstrates their inability to explain the phenomenon of satellites that appear to hover above one spot on Earth. For his part, over the years Sungenis has sought to explain geostationary satellites in various venues and in various ways, both online and in *Galileo Was Wrong*.

After saying that the question put to him "is really an eye-opener," Sungenis explains that "GPS [Global Positioning System] satellites are exactly 22,236 miles above the Earth." This is wrong. That is the altitude of geostationary satellites, but GPS satellites are not among their number. GPS satellites orbit the Earth at much lower altitudes, around 12,550 miles above sea level. They are not geostationary and do not appear as stationary to ground-based observers.

All geostationary satellites orbit at the latitude of the equator. If GPS satellites were confined to the equator, their signals could reach only a portion of the globe. For the GPS system to work, its satellites must be visible from places far removed from the equator. The GPS system relies on a variable number of active satellites—thirty at this writing—the orbital paths of which constantly shift over the face of the Earth. With that many satellites in use, any GPS receiver is able to get signals from several satellites at once, no matter the receiver's location.

Sungenis says that "a GPS satellite has directed to it electromagnetic beams from all over the hemisphere of the

Earth. . . . All these locations depend on the GPS satellite to be perfectly stationary so that it can send back electromagnetic beams with the same precision it received them." This, he says, would not be feasible if a satellite moved in conjunction with a rotating Earth because the satellite would not be able to remain perfectly synchronized with the planet's rotation. Any variance in the satellite's position would make it impossible for it to return signals accurately to terrestrial transceivers. For GPS to work, the satellite must be entirely immobile.

"This means that if the GPS [satellite] either slows down or speeds up by even a fraction, all the signals it sends back to Earth would be off target. In other words, if the GPS suddenly went 6,855 miles per hour instead of 6,856 miles per hour, even for a second or two, the GPS would not send back the signal on target. Suffice to say, there is no known mechanism of man that could keep such precise speeds for any great length of time. And it is a fact that most GPS satellites don't need adjusting for long periods of time."

This means, thinks Sungenis, that GPS satellites must be floating stationary above a rotationless Earth. "The GPS satellite is stationary over the Earth because the Earth is stationary. The GPS satellite doesn't need adjusting very often because there are few things that interfere with its stationary position at 22,236 miles above the Earth. . . . There is a good reason that the only distance a GPS satellite will work is 22,236 miles above the Earth. There is either a gravitational or magnetic force holding it there."

There are multiple problems with Sungenis's argument. The problems are so fundamental that they make a hash of his claim to understand what is going on with satellites.

First of all, GPS satellites are not in geostationary orbits at all. Sungenis is correct to note that geostationary satellites are in orbits 22,236 miles above the surface of the Earth, but those are not GPS satellites, which orbit the Earth at lower altitudes. All geostationary satellites orbit above the equator. Other satellites, including GPS satellites, follow paths that are inclined to the equator. A geostationary satellite orbits the Earth at 6,878 miles per hour. At that speed, and at 22,236 miles above the Earth, the orbit takes exactly one sidereal day, so it appears that the satellite is motionless above a particular spot on the equator. Why are such satellites all at the same altitude? It has to do with the gravity of the Earth and the speed of the Earth's rotation.

A satellite can orbit the Earth at a higher or lower altitude, but it would not keep exact pace with the rotation of the Earth. If it tried to do that at an altitude lower than 22,236 miles, it would need to travel slower than 6,878 miles per hour because it would have less distance to cover in twenty-four hours. The problem is that a satellite trying to keep pace with the Earth at such a lower altitude would not be moving fast enough to experience a centrifugal force strong enough to keep it out of the clutches of Earth's gravity. The satellite would sink toward the Earth.

Conversely, a satellite at an altitude greater than 22,236 miles would have to traverse a longer path than would the standard geostationary satellite. For that higher satellite to attain a geostationary position, it would have to move faster than 6,878 miles per hour, but if it does so the centrifugal force would exceed the pull of gravity, and the satellite would float away from the Earth.

Only at 22,236 miles above the surface are the forces

balanced. This, of course, presumes that the Earth rotates. What if it does not, as in the geocentric system? The most common explanation has been that there is an altitude above the Earth at which the gravitational pull of the distant stars exactly balances that of the Earth. Through a wondrous coincidence, say geocentrists, that altitude also is 22,236 miles above the surface.

All disputants, heliocentric and geocentric, acknowledge that, if the Earth rotates, the heliocentrists' calculations not only are proper but are compelling: the gravity and the rotational speed of the Earth allow only one altitude at which a satellite can appear to be suspended in the sky. There is no such necessity if the Earth does not rotate. The altitude of the equilibrium point—where the pull of the Earth is exactly countered by the pull of the stars—could have turned out to be at any distance above the surface, from a few miles to hundreds of thousands of miles. How odd that the cosmos is so arranged that the stars' pull is precisely enough to keep a satellite floating at 22,236 miles up, just as in the case of a rotating Earth!

His comments show that Sungenis does not understand how GPS satellites work. They send out signals indicating their own position and the time. The signals are broadcast not in pinpoint fashion, as though aimed at a single terrestrial transceiver, but widely and are picked up by GPS receivers that may be hundreds or thousands of miles from one another. The receivers take signals from three or more satellites to calculate their own location. If signals are received from three satellites, the receivers' two-dimensional locations on the surface of the Earth can be determined. If signals are received from at least four satellites, the receivers' elevations also can be determined. The more signals received, the greater the precision of the calculation, down to a matter of a few meters.

Individual GPS receivers, whether handheld (as in mobile phones) or panel-mounted (as in aircraft) or dash-mounted (as in cars) are just that—receivers. They are not transceivers—that is, they do not have transmitters. They receive signals from GPS satellites but do not return signals to the satellites. In the GPS system, the communication is unidirectional (except that the satellites can receive maneuvering instructions from ground control). Thus there is no problem with ground-based GPS transceivers trying to reply to satellites with pinpoint precision or with satellites having to aim their signals at discrete receivers—which indeed would be difficult since GPS satellites move constantly across the sky.

Sungenis has discovered a non-existent problem. He thinks a non-rotating Earth is implied by the near-impossibility of small terrestrial transceivers sending signals to satellites that apparently hover 22,236 miles in the sky while traveling 6,856 (actually, 6,878) miles per hour, but those terrestrial "transceivers" do not transmit at all. GPS receivers on Earth are receivers only, and it is no more difficult for them to pick up signals from overhead satellites than it is for a car's radio to pick up radio signals as the car speeds down the highway.

The bigger problem for Sungenis is not his confusion about how the GPS system works but his inability to explain how geostationary satellites stay suspended above fixed points on Earth. An unnamed questioner asked him to explain how geostationary satellites are placed in orbit and how they remain in position:

> I best understand you by saying that you hypothesize a zone at a certain height from the Earth's surface in which the gravitational force from all bodies cancel each other out or are too weak to matter. . . . so, if you can get a satellite there, then it will stay there. . . .
>
> The most fuel-efficient manner to do this would be to launch straight up with full thrust and then taper (or cut off) while still in the gravity zone so that gravity . . . will cause you to decelerate to no speed by the time you get to the desirable zone. . . . Then you will hover and use stabilizers to orient as you wish. Is this in fact how they get the rocket up there?[97]

The real answer is "No, that is not how they get the rocket up there." Those who launch a geostationary satellite do not intend to decelerate it to zero velocity. They believe they are sending the satellite into orbit at a speed of 6,878 miles per hour, just the right speed for the satellite to match the rotational speed of the Earth if the satellite is positioned 22,236 miles above sea level. Certainly the technicians who control the satellite know whether they are keeping it at that speed or are reducing its speed to zero. The very fact that they make

97. Ibid.

no attempt to stop the satellite's forward motion shows that Sungenis misunderstands the physics of the situation.

This is his reply to the unnamed questioner:

> I don't particularly see the problem you are assuming placing the GPS would entail. [He continues to confuse GPS satellites and geostationary satellites.] Naturally, they would be firing the retro-rockets to stop the ascent. [This is something the technicians in fact do not instruct the satellite to do, because they are not interesting in bringing the satellite's speed to zero.] All they would have to do once it is relatively stable is fire the retros when needed to keep it in the position they need for transmission.
>
> In fact, placing it over a stationary Earth would be much easier than over a rotating Earth. The math is going to be the same whether they think the Earth is rotating or whether it is standing still.

No, the math will not be the same, but, even if it were, what puts the kibosh to Sungenis's understanding is that the technicians make no effort to bring a geostationary satellite to a standstill. They do not fire retro-rockets as he imagines they do. They may fire them to make minor adjustments so the satellite reaches precisely a speed of 6,878 miles per hour, but they do not fire the retro-rockets in order to bring the speed to zero.

Sungenis's argument depends on placing the satellite at a point at which the gravitational attraction of the Earth is exactly countered by the force of something on the other side of the satellite. He says that there "is either a gravitational or magnetic force holding [the satellite] there." We can discount a magnetic force. Satellites are made chiefly of aluminum, titanium, composites, and plastics that are not acted upon by magnetic forces, except extremely weakly. They are not made of iron. The only force Sungenis plausibly can argue for is that of gravity. To keep the satellite from falling to Earth or from floating away, that gravitational force would have to equal the gravitational force exerted by the relatively close and large Earth.

Sungenis believes he has discovered such a powerful force: it emanates from the distant star field. Although stars are immensely far away, and although each individual star, no matter how large, exerts a minuscule gravitational force on the Earth because of its great distance (gravity declines according to the square of the distance), there are so many stars that, cumulatively, they provide sufficient pull to keep a satellite hovering above a spot on the Earth's surface—or so says Sungenis.

The point of equilibrium is found at 22,236 miles above the Earth's surface. At that point—about five times the radius of the Earth—the Earth's gravity is attenuated enough that it is counterbalanced by the gravitational pull of the distant stars. If the satellite were any lower, the Earth's gravity would pull it into the atmosphere and to its destruction. If the satellite were any higher than 22,236 miles, it slowly would move out of Earth's grasp and would float toward the stars, moving directly away from the Earth.

This theory seems almost plausible, until one recognizes that if stars "behind" the satellite exert a gravitational pull, stars "in front of" the satellite and "behind" the Earth exert a pull too. These pulls are nearly equal. They would not be exactly equal because the satellite would be 52,390 miles closer to the stars "behind" it than to those "in front of" it: the radius of the Earth is 3,959 miles, making the distance of the satellite from the Earth's center 26,195 miles (22,236 plus 3,959). Double that to get the total variance.

This variance is insignificant compared to the distance of the stars from Earth. The nearest star, Alpha Centauri, is 26.2 trillion miles from Earth, which makes 52,390 miles less than a rounding error. For practical purposes, the stars "behind" the satellite are as far away as the stars "in front of" it, and the gravitational pull from each direction is the same. That leaves Sungenis's satellite with no invisible means of support. The satellite would have the Earth's tremendous gravity acting on it on one side and an immeasurably small net force from the star field on the other side. It would crash into the Earth.

All this presumes Sungenis's notion of a star field that acts as a uniform shell. He must posit such uniformity if he is to have the star field rotate around the Earth each twenty-four hours. If the star field were not uniform in its gravitational force, over time its varying pulls would throw the planets out of their orbits. But such a uniform star field introduces another and larger problem for Sungenis: the shell theorem. It states that if there is a sphere with a uniform shell (a hollow ball would be an example), then at any point within the sphere the net force of gravity from the shell is zero.

This is easy to imagine if an object is at the center of the sphere; all parts of the shell are equidistant from it. It is harder to visualize for other locations, particularly those near the shell (which, in the geocentric model, means near the stars). One must keep in mind that the closer an object is to the shell, the greater the amount of shell there is "behind" the object. While each part of the shell "behind" it will exert less gravitational pull than will any part near and "in front of" the object, there is much more shell "behind" the object.

The mathematical proof, which was developed by Isaac Newton, is conclusive. When a shell is uniform (even if thick, as geocentrists admit the star field is), at each point within the sphere demarcated by the shell there is no net gravitational pull. Objects remain suspended, not pulled in any direction. This is true, of course, only for isolated objects. If two objects are in proximity—say, the Earth and a satellite—there will be gravitational attraction between them. What the shell theorem really means for Sungenis is that his star field exerts no gravitational pull whatsoever on a geostationary satellite. If he insists on a uniform star field, he does not end up even with the minuscule net pull of the stars that was discussed in the previous paragraphs.

Opponents of geocentrism often say that their most effective argument against the theory concerns geostationary satellites. It is a challenge easily understood by non-scientists. If geocentrists are unable to explain convincingly how a satellite can appear to remain fixed above a particular spot, people who otherwise might give them a hearing turn away. One might think, then, that particular attention would be given by geocentrists to the issue. Not so.

As noted in this chapter's opening paragraph, Gerardus Bouw devotes only one page to the issue in *Geocentricity*. What about Robert Sungenis? The first two volumes of

Galileo Was Wrong: The Church Was Right—the volumes that deal with scientific rather than religious arguments—total 1,379 pages. Geostationary satellites are given less than a single page of text. In that text Sungenis attempts a different explanation of how they work.

He begins by saying, "The heliocentric system explains this phenomenon by viewing the Earth as rotating with a twenty-four-hour period, while the geostationary satellite remains motionless in space. As such, at a specific location on Earth (let's say New York City) one will see the satellite directly overhead one specific time during the day." He again confuses geostationary satellites and geosynchronous satellites. The latter, as suggested by the root of the adjective, are synchronized so that, indeed, they appear over a particular spot at a particular time each day—but they do not remain over that spot. They just come back to it at regular intervals. They move across the sky and do not appear to be stationary.

Geostationary satellites, by contrast, seem to hover perpetually above one location, and that location always is on the equator. It never can be anywhere else. The apparent hovering "works" only because the satellites orbit the Earth with the same angular velocity that the Earth rotates on its axis. It is not possible to have a geostationary satellite hover above any spot that is not on the equator. Contrary to what Sungenis says, New York City is not a candidate because it is not on the equator. No one in the Bronx ever will be able to look up and see a geostationary satellite, though he might see a geosynchronous satellite as it passes over the city on its regular transit.

So why, in the new geocentrists' understanding, does a geostationary satellite remain in apparent suspension above a fixed point on the equator? How can that occur, if the Earth is motionless, not rotating on its axis? In *Galileo Was Wrong* Sungenis does not bring up his notion that the satellite is held in place by the gravitational pull of the stars. He offers a quite different explanation. He says that in fact the satellite is moving swiftly through space at "7,000 miles per hour eastward against the westward rotating universe, which will allow the satellite to remain stationary over a particular location on Earth." The universe, supposedly rotating around the Earth, carries the satellite in one direction, and the satellite flies in the opposite direction at the same speed, resulting in apparent immobility.

It is as though a swimmer were swimming upriver at precisely the speed with which the current seeks to carry him downriver. The result would be that he would remain stationary with respect to a point on the bank of the river. In this analogy is it clear what retards the swimmer's motion: the contrary motion of the medium (water) through which he is trying to swim. What is the situation for the satellite?

It is not, apparently, that the aether seeks to pull the satellite along. That would seem to be the next step in Sungenis's argument. No, he switches gears and says that, in the geocentric arrangement, "the rotating universe generates a real centrifugal force on the satellite" and that this force is "balanced by the gravity of the Earth." Somehow the rotation of the universe pulls the satellite directly away from the Earth. But this explanation will not do. If it is a centrifugal force that keeps the satellite at a certain level, the force can exist only if the satellite itself is moving. Motionless objects are not subject to centrifugal force.

Sungenis's explanation, given in less than a page in

Galileo Was Wrong, is bifurcated—does the satellite fly against a rotating universe, or does a rotating universe pull the satellite upward through centrifugal force?—and is inconsistent within itself. It also is inconsistent with his argument elsewhere about the gravitational pull of the stars. He offers multiple explanations for how a satellite remains suspended above a particular spot on a rotationless Earth, but not one of his explanations will work. This is why he and other geocentrists, such as Gerardus Bouw, devote few pages to a discussion of geostationary satellites. It is to their advantage not to draw anyone's attention to one of the most glaring deficiencies of the geocentric proposition.

MORE WRONG THAN GALILEO

T HE FIRST TWO volumes of *Galileo Was Wrong: The Church Was Right* are subtitled *The Evidence from Modern Science*, and the third volume is subtitled *The Evidence from Church History*. Robert Sungenis and Robert Bennett are listed as co-authors. The work is mainly by Sungenis, but Bennett has a not inconsiderable contribution: a 231-page chapter in the second volume. The three volumes are self-published, apparently through a print-on-demand service. Sungenis designed the simple covers and credits Mark Wyatt for providing the numerous photographs and charts and Rick DeLano for "his helpful research."

Before the work originally was published, it was titled provisionally as *Not by Science Alone: Modern Science at the Crossroads of Divine Revelation*.[98] That title would have been in conformity with Sungenis's earlier apologetical writings: *Not by Faith Alone, Not by Scripture Alone*, and

98. Robert Sungenis, "CAI's Science Creed," (2002), formerly at catholicintl.com/epologetics/scicreed.html.

Not by Bread Alone.[99] In terms of marketing, the actual title is an improvement because it is unexpected. A potential reader is drawn to ask, "Why was Galileo wrong?"

The provisional title, by comparison, lacks any note of the provocative and would be a draw only to those already familiar with Sungenis's apologetical series. His target audience for his geocentrism set is wider than that. He needs a title that will appeal to people uninterested in apologetics, a field in which he no longer much occupies himself. Despite its title, *Galileo Was Wrong* has sold poorly. Perhaps it has been its bulk or its off-putting price. Perhaps it is that not many people are interested in cosmology—or at least not as many as Sungenis expected.

In the first volume, immediately following "About the Authors," are thirteen endorsements. Among the endorsers are Gerardus Bouw, the Protestant author of *Geocentricity*, the next-largest pro-geocentrism work; E. Michael Jones, editor of *Culture Wars* magazine, who was a speaker at the 2010 geocentrism conference; Thaddeus Kozinski, a philosophy professor at Wyoming Catholic College who has opined that the 2012 Sandy Hook Elementary School shootings were faked and that the Twin Towers were brought down by "a controlled demolition";[100] John Salza,

99. Robert Sungenis, *Not by Faith Alone* (Goleta, CA: Queenship Publishing, 1997); *Not by Scripture Alone* (Goleta, CA: Queenship Publishing, 1998), *Not by Bread Alone* (Goleta, CA: Queenship Publishing, 2000).
100. Like Robert Sungenis, Thaddeus Kozinski subscribes to multiple conspiracy theories. In an online exchange with professor of philosophy Edward Feser, Kozinski gratuitously injected comments that were off topic. As Feser summarized it: "On the basis of a passing reference someone made about Sungenis, you suddenly went off on a series of rants about the 'official narrative' regarding the Holocaust, 9/11, [World Trade Center] 7, Gaza,

who in *The Remnant* has written that Protestants cannot be saved; and "Anonymous," who says, "Here you will find a thorough review of the scientific observations along with a review of the scientists themselves."

This last comment is worth noting because portions of *Galileo Was Wrong* are animadversions about the personal and intellectual lives of men such as Albert Einstein and Johannes Kepler. Sungenis believes it is proper to consider not just a scientist's writings on science but also the conduct of his personal life and his views on other matters (presumably including any opinions he may have about conspiracies).

To an extent this is true, and to an extent it is false. It is true that a scientist's ideas on matters other than science may tell us something about his ability to reason well. It is true if it helps us judge whether he has the capacity to draw proper conclusions from evidence at hand. (This applies to Sungenis himself, which is why his writings on Jews and other issues are worth noting: do those writings suggest he is capable of making well-balanced judgments?) But Sungenis is wrong to argue that much can be inferred about the worth of a scientist's scientific work from an examination of his personal life. There is no obvious connection between a man's moral failings, for example, and his capacity to reason scientifically.

Sungenis, using the editorial first-person plural, explains in a "Notice Concerning Terminology and Physics"

the Iraq war, Afghanistan, Sandy Hook, Likudniks, the military-industrial complex, the [Anti-Defamation League], 'useful-idiot Catholics' doing 'the regime's' bidding, etc. etc." (April 5, 2013), edwardfeser.blogspot.com/2013/04/reply-to-kozinski.html.

that "[w]e employ the term 'geocentrism' to represent the scientific position that the Earth is motionless in space at the center of the universe with neither diurnal rotation nor translational movement." This is how most geocentrists use the term, though some posit an Earth that rotates on its axis while nevertheless lying at the center of the universe. (Among other things, this lets such geocentrists get around the problem of how the entire star field could circle the Earth each twenty-four hours.)

Sungenis notes that "[o]thers employ 'geocentricity' or 'geostatism' to represent the motionless Earth and employ 'geokineticism' or 'antigeostatism' to represent a moving Earth. The term 'geocentrism' will stand for any scientific theory that holds the Earth is the center of the universe and/or motionless in space." ("Geocentricity" is a term that Gerardus Bouw claims to have invented.) In contrast, says Sungenis, "'heliocentrism' will stand for any scientific theory that holds that the Earth is not in the center, or that the Sun is in the center, or that there is no center of the universe, and that the Earth is in constant motion."

In a short introduction, last updated at the end of 2012, Sungenis says, "Unbeknownst to almost the entire human race, however, is the fact that no one in all of history has ever *proven* that the Earth moves in space. Despite his protestations to the contrary, the historical record reveals that Galileo Galilei had no proof for his controversial assertions. What he purported as proof in his day would be laughed out of science classrooms today. Galileo merely began a myth, a myth that eventually took on a life of its own and became the status quo of popular thinking." Sungenis cites Albert Einstein, Stephen Hawking, and

Arthur Eddington to the effect that it is not possible for science—particularly since the adoption of the theories of relativity—to determine whether the Earth is at absolute rest or in motion.

The same can be said of any celestial object. If it is even possible, under relativity, to argue that the universe has a center and that the Earth is at rest at it, then it equally is possible to say that any other object, whether a planet or a star or a speck of intergalactic dust, is motionless at the center. There is no way to know.

There is a further problem. It is one thing to say that there exists a point in space that is the center of the universe. It is something else to locate that point (again, assuming such a point exists; it may not, under current scientific theories) and to determine whether anything other than empty space is there. Before one can determine what is located at that central point, the central point itself has to be found. No geocentrist has offered a means to do this, beyond the gratuitous assumption that Earth is at the center. There has been no indication of how someone finding himself at a distant spot in the universe might determine where its center is.

It may not be apparent immediately to someone coming across *Galileo Was Wrong* or other pro-geocentrist works for the first time why it should make a difference whether that central point is occupied by the Earth. Would it not be enough, for geocentrism to be true, to allege that the Sun and the planets revolve around the Earth? That would result in an Earth system instead of a solar system. The system would be geo-centered. To the minds of today's geocentrists, that would not be enough.

An Earth-centered system that moves through space would be, to use Walker Percy's phrase, "lost in the cosmos." It would not be special. It would not be significant. It would not have the prestige that comes with being located at dead center, and it is this prestige that the new geocentrists are particularly intent on maintaining. They think Christianity's claims would be in jeopardy if God did not position Earth at the center of all things. They think man's status is as much a consequence of his location as of God's supervening care for him. This is the undergirding notion in *Galileo Was Wrong*. It is the notion that determines how the book's chief author interprets and, not uncommonly, skews the writings and work of the hundreds of people he cites.

"Knowledge is plentiful," writes Sungenis, "but wisdom is severely lacking. As one astronomer has admitted: 'Perhaps it is time for astronomers to pause and wonder whether they know too much and understand too little'":

> Hence, the first two volumes of *Galileo Was Wrong: The Church Was Right* will be devoted mainly to the scientific evidence concerning cosmology. Since modern science has made itself into such an imposing authority on the minds of men today, no study of this kind could possibly be adequate until the scientific assertions are thoroughly addressed and rebutted. We have compiled the most comprehensive scientific treatise on the issue ever offered to the public. The third volume will be devoted mainly to the scriptural, ecclesiastical, and patristic evidence supporting the cosmology of geocentrism.

Sungenis realizes that many people will approach his writings and those of other geocentrists with suspicion:

> We only ask that you, the reader, contemplate the issue with an open mind. All too often when controversial subjects of this nature arise, those who wish to protect the status quo are quick to demonize their opponents, choosing instead to associate them with such institutions as the Flat Earth Society or characterize them as geeks who don tinfoil hats and receive messages from outer space. Hopefully, you will not fall into that trap of bigotry and censorship. Rest assured, the authors of this book do not fill any of the above caricatures.

It is true that neither Robert Sungenis nor Robert Bennett belongs to the Flat Earth Society (an actual organization founded in 1956), and the same can be said for each of the geocentrists mentioned in *The New Geocentrists*. They have not been seen to "don tinfoil hats," and they have not claimed to "receive messages from outer space." No heliocentrist has asserted that they have. But some of them—Sungenis in particular—subscribe to such outlandish ideas regarding other matters that their competence in judging evidence of any sort is called into question.

Is it prudent for a reader who has no formal scientific background to rely on the writings of someone who happily boasts that he is upending the scientific establishment, particularly if that writer has shown himself elsewhere to be incapable of making reasoned distinctions and of evaluating evidence at arm's length? Is it prudent for a lay-man to rely on a writer who has subscribed to multiple and often ludicrous conspiracy theories? Is it prudent for him to ignore four centuries of scientific investigation (much of it by Christians who were as sincere in their faith as are today's geocentrists) and to adopt the ideas of a writer who has no formal training in science beyond a few lower-division college courses? Is it prudent for the reader to take as his own the historical and theological judgments of some-one who is neither a historian nor a theologian?

The new geocentrists say "yes" to all of these questions. They are confident of their skills and of their insights. They believe they have come across truths that have eluded intel-ligent investigators for centuries or that, more commonly, have been hidden away deliberately, the better to foster among the general population the irreligion of the scien-tific chattering classes. In their books and articles the new geocentrists seek to share their discoveries with people at large—not with an expectation that most readers will have a revelatory experience and end up as devoted follow-ers but with a hope that they might "realize that there is enough evidence supporting geocentrism to cause a rea-sonable doubt in the minds of intelligent people." That, says Sungenis, is the "modest goal" of *Galileo Was Wrong*.

Herbert Thurston, an English Jesuit, once commented that he could find, in any ten pages by American historian Henry Charles Lea, an average of one blunder per page. Thurston was challenged to do so by G. G. Coulton, a pro-fessor of history at Cambridge who considered himself to be Lea's English counterpart. Coulton selected ten pages from Lea at random, added two more for good measure, and defied Thurston to find ten blunders in the twelve pages. Thurston found fifteen.

That episode from the 1930s comes to mind when I page through *Galileo Was Wrong*. I have not had the patience to read the three volumes completely (I wonder whether anyone beyond the author's entourage has read through the 2,200 pages from first to last), but the substantial chunks I have read lead me to think that, were he alive, Herbert Thurston could claim of *Galileo Was Wrong* what he claimed of Lea's writings.

Someone with no other call on his time and with the patience of Job could dispute Sungenis's arguments page by page, for there is plenty to dispute. I have yet to open the text at random and not find a wrong emphasis or a false inference. While much of the reportage is accurate, often the interpretation is not. In many places evidence has been cherry-picked. Conclusions sometimes fail to follow from premises, and often an alternate (and usually more obvious) conclusion simply is not mentioned.

As one might expect, the long chapter by Robert Bennett, who holds a Ph.D. in physics, contains a fair amount of mathematical formulation. Curiously, there are more equations and more difficult mathematics in the portions of the book written by Sungenis. He has taken the more convoluted equations from elsewhere and has given appropriate citations, but there is no indication that he understands the equations or could "do" them himself. Nevertheless, he gives the unwary reader the impression than he knows more physics and mathematics than does the reader and perhaps more than professional scientists.

The second chapter of *Galileo Was Wrong* is titled

"Answering Common Objections to Geocentrism." There are twenty-nine subparts. Most of them are worded in ways sufficiently colloquial that readers should have little difficulty in following Sungenis's argument, but some subparts are thick with equations. Are the equations correctly written? Does one equation lead properly to the next? Almost no one who is likely to purchase the text would be able to answer those questions. The value of the many complex equations is not in the way they prove to the reader (or to the author) the truth of a statement but in the way their sheer weight seems to give a scientific imprimatur to *Galileo Was Wrong*.

The twenty-nine subparts in the second chapter cover such topics as stellar parallax, Foucault's pendulum, geosynchronous satellites (in one page, with much confusion, as noted above), whether the four seasons prove the Earth rotates, and the implications of stars' red shift. The following chapter has twenty-two subparts, each offered as evidence for the Earth being at the center of the universe. The fourth chapter has twenty-one subparts, all dealing with Earth's supposed motionlessness.

In each of these subparts, no matter the immediate topic, Sungenis believes he presents, if not a fully convincing argument, at least an argument that points toward geocentrism. He seems to think that the cumulative effect of these seventy-two subparts is enough to convince the open-minded reader. These three chapters form the core of his argument. The remaining chapters in the first two volumes add supplementary material that is not key to the basic question: "Is geocentrism true?"

The second volume, aside from including Bennett's

long chapter, has chapters about "The Cause of Gravity in the Geocentric Universe," "How Old and How Big is the Geocentric Universe?", and the response of modern science to the geocentric argument. The final chapter of that volume contains nine subparts, the titles of which leave little doubt what Sungenis thinks about each topic. Among them are "Is Modern Science Corrupt?", "Nicolaus Copernicus: Too Many Pagan Influences," "Johannes Kepler: Suspected of Murdering Tycho Brahe," and "Isaac Newton: Climbing the Ladder of Success."

Throughout *Galileo Was Wrong* Sungenis questions the motives, backgrounds, and private lives of scientists associated with the heliocentric position. He does this not just for scientists who lived centuries ago but for those of recent times—Einstein, for example.

The problem is not that such questioning consists of *ad hominem* attacks. Usually it does not. Sungenis does not argue that a scientist's argument was wrong because he was a bad man. He argues that a scientist's personal life and his thinking on other matters (religion, for example) may have influenced the way he approached scientific issues. It is fair, he says, to ask whether prejudices on non-scientific matters held by Copernicus or Kepler or Newton ended up coloring their thinking, and it equally is fair to ask whether modern scientists operate from social or cultural or political presuppositions that limit their ability to judge objectively.

What is at stake is not just the reliability of scientists' arguments (can they "do" physics, astronomy, and mathematics well?) but their reliability as authorities. Can their representations be relied on not just by non-scientists but by other scientists? Has it been prudent, over the centuries, to repose trust in Copernicus, Kepler, and Newton, and is it prudent today to repose trust in Einstein, Hubble, and Hawking?

However one judges the reliability of these scientists, whether of earlier centuries or of our own, the same approach can be applied toward Sungenis himself. He should not object to seeing his own background and non-scientific writings brought up, his own prejudices examined, his own predilections discussed—not because those things speak directly to his scientific arguments (they rarely do) but because they speak to his capacity to judge.

His obsessive anti-Jewish writings—hundreds of thousands of words spread over more than a decade—may be irrelevant to his calculation of the gravitational force exerted by the Sun on the Earth, but those writings are not irrelevant when it comes to weighing his overall judiciousness and his ability to be fair to scientists with whom he disagrees (he does not hesitate to note that some scientists he disagrees with were Jewish). Sungenis's anti-Jewish writings may be a signal to readers of *Galileo Was Wrong* and his other pro-geocentrism books that they need to look critically at each representation he makes and not assume that he necessarily plays fair with his scientific opponents. If Sungenis spent a decade accusing Jews of innumerable conspiracies against common sense and the common good, are his representations about supposedly nefarious scientists to be taken at face value?

Galileo Was Wrong should not be laid aside without reference to its third volume, which has four chapters. The first concerns "Scripture Passages Teaching Geocentrism." The next is about "The Consensus of Church Fathers." Then comes

"The Catholic Church's Teaching on Geocentrism." The final chapter of this volume is on "Interpreting Genesis 1."

Protestant geocentrists will be most at home with the first of these chapters. About thirty scriptural passages are discussed in the course of forty pages. Some of those passages were considered in earlier chapters of *The New Geocentrists*: 2 Kings 20:9-11, Joshua 10:12-14 ("Joshua's long day"), Ecclesiastes 1:5, and Psalms 19, 93, and 104.

In not one of the passages does Scripture actually teach that the Sun revolves around the Earth, and not a single verse says that the Earth is at the center of the universe—a fact that Sungenis does not advert to. Each passage uses phenomenological language to describe what is seen. This is done even today by professional astronomers, who unanimously accept heliocentrism, when they refer to the Sun rising and setting. In using such language, the astronomers deceive no one, neither themselves nor others.

It is no more wrong for an astronomer to talk about the Sun setting than for a poet, standing on a western shore, to write that the Sun sinks beneath the water. No present-day poet believes, as some of the ancients believed, that the Sun's fires are quenched each evening in the sea, only to be reignited the following morning, but it is not wrong for a poet to employ such figurative language. So with Scripture, which throughout uses analogies, similes, and metaphors. (Many examples can be found in the Song of Songs and in the Psalms.)

Sungenis demands a literalistic interpretation of the passages he lists. To an inquirer he wrote, "Church documents only allow for the figurative interpretation if the literal cannot be fulfilled. No Church document allows for a carte blanche figurative interpretation of Scripture. The Church has always held that the literal comes first. Leo XIII was very clear about that in *Providentissimus Deus*." A little further on, he expands his thought: "[Y]ou haven't proven that the cosmological Scriptures are figurative. If you can't, then you are bound to take it [sic] literally."[101]

This is not what is taught in *Providentissimus Deus* nor in the later *Divino Afflante Spiritu*. Neither papal document insists that a reader must try to shoehorn a literalistic interpretation into Scripture. In *Providentissimus* 19, for example, Leo XIII says, "The unshrinking defense of the Holy Scripture, however, does not require that we should equally uphold all the opinions which each of the Fathers or the more recent interpreters have put forth in explaining it; for it may be that, in commenting on passages where physical matters occur, they have sometimes expressed the ideas of their own times, and thus made statements which in these days have been abandoned as incorrect."

Using Sungenis's criterion—that a figurative interpretation can be used only "if the literal cannot be fulfilled"— leaves one wondering whether any passage then is subject to a figurative understanding. Sometimes a Protestant will protest against the Catholic Church's literal interpretation of "This is my body" (Luke 22:19) by pointing out that elsewhere Christ said he was a door or gate (John 10:9). No one thinks he actually became a door or a gate, so why should one think he takes on the appearance of bread?

I dealt with this and similar arguments in *Catholicism*

101. Robert Sungenis, "The Fallen Star: Relativity Meets the Absolute" (2002), formerly at catholicintl.com/epologetics/geofallenstar.html.

and Fundamentalism, where I showed that the situations were not parallel and that it is fairly easy to determine when a scriptural passage ought to be taken figuratively. The argument I used was not of my own devising; it had been the common property of apologists for centuries. It did not presuppose Sungenis's over-strict understanding of when a figurative usage is permitted or that one is "bound to take" a passage literally unless there is no possible literal way to understand it.

As he misapplies the verses he cites, so Sungenis misapplies the many quotations he gives from the Fathers of the Church, those orthodox writers of the first centuries. He gives dozens of quotations from them, dividing the quotations into those referring to cosmology in general, to the "length of the Genesis day," to the definition of the firmament, and to the spherical shape of the Earth.

Opponents of geocentrism acknowledge that early Christian writers accepted a geocentric view of the cosmos. In those centuries, nearly everyone did. Such a view formed a backdrop to the Fathers' writings, but nowhere did they argue for the scientific factuality of geocentrism over heliocentrism. This is seen (or not seen) in the sixty pages of quotations served up by Sungenis. Did the Fathers demonstrate a consensus of opinion regarding geocentrism? Yes and no.

They all wrote as though geocentrism were an accurate explanation of what they saw around them, but they did not write in favor of that theory over against a competing theory. Astronomy was not their concern, and they made no attempt to teach it. Similarly, they accepted the ancient understanding that the visible world is comprised of four "elements"—earth, water, air, and fire—but nowhere did they teach that this was accurate science. The Fathers accepted also the notion of the four humors—sanguine, choleric, melancholic, phlegmatic—that in Greek thinking were the four basic personality types, but they nowhere attempted to define these as accurate psychology.

In the thirteen-page introduction to the third volume, Sungenis opens by saying, "If you have read the first two volumes of *Galileo Was Wrong: The Church Was Right*, you are now ready to read volume three. As was the case with the science, the historical issues concerning Galileo shows [sic] that the data is plentiful but the correct interpretation is almost always lacking." Sungenis proposes to give the correct interpretation. He does not make it easy for the reader to conclude that he is capable of doing so.

He refers to Ronald W. Clark's biography of Albert Einstein, *Einstein: The Life and Times*.[102] In the introduction Sungenis photo-reproduces two pages from Clark's book. On the left-hand image he boxes several sentences. On the right-hand image he underlines the first three lines of text. He presents Clark's words this way:

In the United States Albert Michelson and Edward Morley had performed an experiment which confronted scientists with an appalling choice . . . leaving science with the alternatives of tossing aside the key which had helped to explain the phenomena of electricity, magnetism, and light [or] of deciding that the Earth was not in fact moving at all. . . . For

102. Ronald W. Clark, *Einstein: The Life and Times* (New York: William Morrow, 2007).

there seemed to be only three alternatives. The first was that the Earth was standing still, which meant scuttling the whole Copernican theory and was *unthinkable*.[103]

Several observations need to be made here. Sungenis italicized the final word in the material he quotes, but it is not italicized in the original. That is a minor point. The ellipses are not so minor. Two ellipses are within the quotation, but another might be said to be at the end, after "unthinkable." Sungenis truncates Clark's words to suit his own needs. Clark mentions "three alternatives," but Sungenis lets his reader see only one. Here are the other two, which immediately follow "unthinkable":

The second [alternative] was that the aether was carried along by the Earth in its passage through space, a possibility which already had been ruled out to the satisfaction of the scientific community by a number of experiments, notably those of the English astronomer James Bradley. The third solution was that the aether simply did not exist, which to many nineteenth-century scientists was equivalent to scrapping current views of light, electricity, and magnetism, and starting again.

By ending the quotation at "unthinkable," Sungenis starts to lay the framework for what he regards as a

103. Robert Sungenis, *Galileo Was Wrong: The Church Was Right*, tenth ed., vol. III (State Line, PA: Catholic Apologetics International, 2014), 3.

century's worth of scientific malfeasance. The Michelson-Morley experiment had failed to show that the Earth moved through the aether. Subsequent attempts to re-enact the experiment fared no better. This meant, says Sungenis, that the Earth is motionless. This was "unthinkable" to scientists because a motionless Earth would be interpreted as affirming Christianity, something scientists (nearly all of them supposedly irreligious) sought to avoid at all costs. They were left with no option but to enter into a massive conspiracy to hide the truth from the rest of mankind. (At the 2010 geocentrism conference, Sungenis's first talk was titled "Geocentrism: They Know It But They're Hiding It.")

In reality it is Sungenis who is hiding the truth. He omits two of Clark's three alternatives and does not present the reader with what turned out to be the correct interpretation of the results of the Michelson-Morley experiment: that aether does not exist and, as a consequence, that nineteenth-century "views of light, electricity, and magnetism" were flawed and had to be reworked.

So much for the "ellipsis" at the end of the quotation as presented by Sungenis. The two internal ellipses also deserve examination. The first replaces these words: "Designed to show the existence of the aether, at that time considered essential, it had yielded a null result." This description shows the purpose of the Michelson-Morley experiment. It was intended to prove that aether exists, but it "yielded a null result." At the time it was conducted, in 1887, the experiment was thought to be a failure, but that was only because scientists of the era assumed the existence of aether. In fact, the experiment can be considered a success in that it cleared the field and gave scientists an impetus to look for answers elsewhere.

Although the result of the Michelson-Morley experiment no doubt was disturbing to scientists who had staked their reputations on the existence of aether, aether's non-existence did not put in jeopardy the larger part of their studies. Eventually they found alternate ways to explain phenomena. But the non-existence of aether is fatal to geocentrism—or at least to the majority school of thought within geocentrism, the one holding that the Earth not only is at the center of the universe but also does not rotate on its axis. Without aether, there is no mechanism for the diurnal revolution of the universe around the Earth.

Through his use of the first ellipsis within the quotation, Sungenis tries to keep the reader away from considering that the non-existence of aether is a possibility. He also fosters a wrong understanding of what the Michelson-Morley experiment was intended to prove. Scientists of the time postulated that aether was a substance so thin that its existence was not easily demonstrated. It was not like the wind, which is invisible but can be felt. The aether would not be able to be seen or measured directly through any scientific instrument, but its existence could be inferred if it disrupted the passage of light, even if only minutely. That disruption is what the experiment failed to show.

That absence of a disruption constituted the "null result" of the experiment, which never was designed to prove that the Earth does not move. Scientists of the time were convinced that it does move—the discovery of stellar parallax earlier in the century and of stellar aberration in 1728 were sufficient proof—but they wanted to show the existence of a medium that could transmit light. That medium was the aether. Once they accepted the non-existence of aether, they found other ways to explain what light is and how it traverses space.

What about the second internal ellipsis in the quotation given by Sungenis? It may be the most problematic of the three. In normal usage, an ellipsis indicates that several words and perhaps even a sentence or two have been omitted, without changing the meaning of the passage. In this case, Sungenis has omitted no fewer than 18,000 words. The first part of the quotation, up through the words "moving at all," appears on page 57 of Clark's book. The remaining words appear on page 110. Sungenis has reached back 53 pages to cobble together an authoritative-looking quotation that misleads the reader.

All this is found in the introduction to Sungenis's volume concerning "The Evidence from Church History." It is not an auspicious start. A few pages later—still in the introduction—comes a complaint about Vatican II. That council's

> pro-Galileo mentality led to a complete revamping of how the Catholic Church understood herself and her scriptural foundation, which, incidentally, began in the mid-1800s immediately after Gregory XVI took Galileo's book off the Index in 1835.

> The new view of the Church and Scripture was officially endorsed by Pius XII's 1943 encyclical *Divino Afflante Spiritu* and ended with Vatican II's *Dei Verbum* 11, which, as the modern prelature

desired to understand it in light of the Church's alleged mistake with Galileo, taught the unprecedented idea that Scripture is only inerrant when it speaks on things concerning salvation, not history or science.

Galileo's name nowhere appears in the documents of Vatican II, and the council did not take up the disposition of his writings or Gregory XVI's decision to remove them from the Index. How, then, is the "pro-Galileo mentality" of the council fathers to be understood? It was not as though the bishops were coming to Rome in 1962 to reconsider the Galileo case. If the term means anything, it means that the bishops agreed with Galileo that the Earth orbits the Sun, though there is no evidence that anyone polled them on the issue.

In the third volume of *Galileo Was Wrong* Sungenis builds a four-stage argument. The first stage is that Scripture teaches geocentrism with sufficient clarity that no Christian should miss the point. The second is that the ancient Fathers of the Church unanimously endorsed geocentrism and never came out in favor of heliocentrism. The third is that, as the book's title has it, *Galileo Was Wrong* and the Church was right to condemn his writings—and that the condemnation of heliocentrism was an infallible decision of the Church. The fourth stage of the argument is that recent Church leaders, particularly several popes over the last two centuries, have acted or taught against that infallible decision and have thrown the Church into disarray while not being able to overturn the irreformable earlier teaching.

Before Sungenis gets to those considerations, he has a complaint to lodge against "conservative Protestants" who, like most Catholics, "are prone to biased influence from modern academia. Although some Protestants have forged a valiant fight against evolution by employing the tenets and experimental data of science itself, when issues of cosmology arise, most Protestants invariably side with Darwin's intellectual cousins"—that is, with the Copernicans.

Why do they do that? It is "for the same reason that Catholics do—it is much too embarrassing in today's world to take a strict literal view of the Bible and believe the Earth was made first, is motionless, and was placed in the center of the universe." Sungenis then quotes Danny Faulkner, whose criticism of Gerardus Bouw is given in an earlier chapter of *The New Geocentrists*: "While geocentrists are well intended, their presence among recent creationists produces an easy object of ridicule by our critics." (Faulkner, a Protestant, is himself a creationist.)

Sungenis says that the root of the problem, for both Catholics and Protestants, is that they

> have been unduly influenced by men in white lab coats who write all kinds of fancy equations yet also provide fantastic machines that benefit mankind. Hence, the scientists have convinced the religionists that the scientists know better and that it would be foolish to argue against their theories and equations. . . . But as we have shown in the first two

volumes, once one puts his mind and will to work, it is rather easy to blow down the house of cards that modern cosmology and cosmogony has [sic] built for itself [sic].

An examination of Scripture and early Christian writings, says Sungenis, will show that "geocentrism was, and still is, the Church's official teaching on cosmology. It began with the Church Fathers and was handed down to the medievals, through the Tridentine catechism, and capped by the diligent work and permission of two popes, Paul V and Urban VIII, who dealt directly with Galileo."

Sungenis claims the teaching not only has been clear and direct but saw no opposition until the nineteenth century. "We will also find in our historical study that the two instances in which the Church seemed to relax some of its earlier condemnations against heliocentrism, namely the issuance of an imprimatur in 1820 to Canon Settele's book on heliocentrism, and the removal of Galileo's name from the *Index of Forbidden Books* in 1835 in the reign of Gregory XVI, are instances filled with ecclesiastical malfeasance, and which, in spite of it all, do nothing to change the tradition and decrees of 1616 and 1633 when Galileo and heliocentrism were condemned."

Giuseppe Settele's *Elements of Astronomy* argued that heliocentrism had been proven. The Church's chief censor refused to license the book, and Settele appealed directly to the pope, Pius VII. The Congregation of the Index was instructed to reconsider the matter, and it overturned the censor's decision. The decree allowing publication stated that "His Holiness has decreed that no obstacles exist for those who sustain Copernicus's affirmation regarding the Earth's movement in the manner in which it is affirmed today, even by Catholic authors."

The Pope went beyond mere permission. "[Pius VII] has, moreover, suggested the insertion of several notations into this work, aimed at demonstrating that the above mentioned affirmation [of heliocentrism by Copernicus], as it has come to be understood, does not present any difficulties, difficulties that existed in times past, prior to the subsequent astronomical observations that have now occurred." When the next edition of the *Index* was published in 1835, Galileo's works were omitted from it.

The question of the *Index* and the juridical force of the decision in the Galileo case were addressed in the chapter about Paula Haigh. They need not detain us here, except to note that the placing of a book on the *Index* never has been considered an exercise in papal infallibility, even when a pope explicitly has endorsed such a move.

The same can be said of juridical decisions, such as the one taken by the investigating commission in the Galileo case. A pope might endorse such a decision and believe it both just and factually accurate, and he might convey his endorsement in a bull or other document, but that raises neither the decision nor the pope's action to the level of infallibility. The decision of the authorities in the Galileo case, and Urban VIII's subsequent approbation of the decision, were exercises not in defining doctrine but in discipline. (The same can be said of the decisions of 1820 and 1835.)

This is made clear in the discussion of the Galileo case in the *Catholic Encyclopedia*—published a century ago, in 1914—which says that it "is undeniable that the ecclesiastical authorities committed a grave and deplorable error" when they condemned Copernicanism and that they "sanctioned an altogether false principle as to the proper use of Scripture." This false principle is one that Sungenis persists in using. But what about the seventeenth-century popes, Paul V and Urban VIII, who affirmed the decision of those who oversaw the Galileo case? The encyclopedia article says,

> The question is, however, whether these pontiffs condemned [heliocentrism] *ex cathedra*. This, it is clear, they never did. . . . Nor is the case altered by the fact that [Urban VIII] approved the congregation's decision *in forma communi*, that is to say, to the extent needful for the purpose intended, namely to prohibit the circulation of writings that were judged harmful. The pope and his successors may have been wrong in such a judgment, but this does not alter the character of the pronouncement, or convert it into a decree *ex cathedra*.

Disciplinary decisions can be modified or even reversed. Sungenis sees a great danger in this: "[I]f the popes of the seventeenth century who approved the condemnations against heliocentrism could err, then current popes who approve the reigning opinions of modern science can also err."

He thinks the greatest error regarding these matters came from John Paul II, who in 1992, in an address to the Pontifical Academy of Science, exonerated Galileo. This put the Church in an untenable position, thinks Sungenis. The Pope "save[d] the modern Church on one count," but he then had "the insurmountable problem of explaining how the Church of the past, which claimed to be guided by the Holy Spirit just as much as the Church of the present, could have been duped into thinking that true cosmology was even addressed by Scripture, much less erroneously concluding that the Sun revolved around the Earth."

STUMPED AT A DEBATE

O N MARCH 29, 2011, Robert Sungenis gave a lecture about geocentrism at the University of Manitoba. There were twenty-nine people in the audience. A student who attended reported that Sungenis's "logic seems to come down to mistaking correlation for causation in the downfall of the Catholic Church. It is as follows: People use Galileo as an example of things that the Catholic Church has gotten wrong in the past. Since Galileo's time, the Church has fallen in prominence and atheism has gained in popularity. Ipso facto, heliocentrism leads to atheism."

From the account at her blog, the student certainly did not arrive at the debate predisposed to agree with Sungenis, so there is some inexactitude in her characterization of his position—but not much. He frequently has argued that the Galileo case was not decided in error and that the fortunes of the Church would be much improved if that were appreciated widely.

The student—identified at her blog only as Flora—added:

[Sungenis] continued to say that current theories

based on heliocentric models have not been proven. This is a familiar creationist claim that has been so thoroughly debunked that it's almost tiresome to mention it. He, either deliberately or through some vast oversight in his research, fails to understand that theory cannot ever be 100% proven. The [heliocentric] theory only works in every conceivable situation we have applied it to—and there are a great deal of those!

Of course, what he asks scientists to provide him with are absolute certainties, and, being good scientists, they give him assertions with qualifications. He interprets this as uncertainty and dissent, when in reality it's intellectual honesty.[104]

Flora was not impressed with the massiveness of *Galileo Was Wrong*. She drew an unflattering conclusion:

He shamelessly promoted this book throughout the lecture, claiming to be holding back valuable evidence in support of his ideas. The thing itself could be used to hold down a helium balloon in a hurricane, though at the $80 price tag I would suggest finding a moderately sized boulder instead. He obtained his Ph.D. from an unaccredited distance-education program and is quite proud of the fact

104. "In Which the Universe Revolves Around Robert Sungenis, Part I" (Mar. 21, 2011), subspecies.wordpress.com/2011/03/31/in-which-the-universe-revolves-around-the-catholic-church-part-1.

that his doctoral dissertation [which became the first edition of *Galileo Was Wrong*] is over 700 pages long. By contrast, normal research-based theses are around 150-200 pages long.

Not only does that indicate the sort of quality of education Dr. Sungenis received, it is a lovely demonstration of his complete inability to get to the point.

During his lecture Sungenis spoke about multiple topics, such as the cosmic microwave background, Newtonian mechanics, and alleged errors made by Newton and Einstein. Flora thought that the science was presented poorly. "It was a 'Look! Science! I'm smarter than you, so you couldn't possibly understand this, but trust me, this is science!' kind of moment. He threw around words like 'quasar,' 'isotropic,' and 'anisotropic' without definition or explanation. I was annoyed."

She thought that some of his proofs made little sense. To demonstrate that the Earth is at the center of the universe, he pointed to the Sloan Digital Sky Survey, which, said Flora, "showed all the galaxies in the observable universe, with us at the center. By definition, if we can see a specific distance all the way around us, we will be in the center." If an equivalent sky survey were taken from a planet orbiting Alpha Centauri, that planet would appear to be at the center of the visible universe because, by definition, an observer is at the center of what can be observed. Flora saw the error in Sungenis's argument immediately and wondered why he did not.

The next day came a debate. Flora said she particularly looked forward to it:

> I must confess that I was quite excited by the prospect of having someone debate him. He clearly was an experienced orator, but it was hair-pullingly aggravating to have to sit through a solid hour of his verbal diarrhea [at the previous night's lecture]. The prospect of someone calling him out on his insane conclusions delighted me, though I must admit that I had trepidations as well. I knew in advance that the individual who had stepped forward to debate Dr. Sungenis was not a professor of astronomy or even a graduate of that program. They had recruited, with a week's notice, an undergraduate student.[105]

He was Adam Cousins, whom Flora called "knowledgeable," but she was disappointed that no faculty member volunteered to square off against Sungenis. "I would have sorely loved to see him verbally eviscerated." The audience, which was larger than for the previous day's lecture, included many people not from the university. At the conclusion of the debate a vote was taken, and Sungenis won twenty-four to eighteen.

The topic was "A geocentric system is a false cosmological assumption." Flora thought this put Cousins at a disadvantage. He, a non-debater, would have to take the

105. "In Which the Universe Revolves Around Robert Sungenis, Par II" (Apr. 29, 2011), subspecies.wordpress.com/2011/04/29/in-which-the-universe-revolves-around-robert-sungenis-part-2.

affirmative against an experienced debater, and so long as "Sungenis could plant some seed of a doubt, demonstrate that in some minute way that geocentrism was possible, the debate was his to win." She thought Cousins opened the debate well—by using a prop.

Sungenis believes that the Earth is kept motionless at the center of the universe because the aether rapidly rotates around it, exerting equal force from all sides. Cousins, to demonstrate that the Earth could not remain motionless if the aether (assuming it existed) circled it each twenty-four hours, brought out a dish of water in which a ping pong ball floated. The ball was the Earth, and the water was the aether. He stirred the water in a circle. Instead of remaining motionless at the center of the bowl, the ball started to spin—just what the Earth would do if the aether that abutted it spun in a circle.

Later Cousins argued that the Michelson-Morley experiment and its replications since 1887 have failed to demonstrate the existence of aether. Sungenis held that the experiment instead showed that aether exists and that the Earth does not rotate on its axis—a conclusion not shared by the large majority of physicists. Sungenis and Gerardus Bouw maintain that aether is the substance through and by which light supposedly is propagated. They and other geocentrists say that the famously "failed" Michelson-Morley experiment (the most famous thing about it is that it "failed") argues in favor of their position in that it failed to demonstrate the motion of the Earth. Its actual purpose was to prove the existence of aether, and that is what it failed to do.

Then it was Sungenis's turn. Flora was not impressed by

what she heard. "He argued that the mass of the universe isn't accounted for by heliocentrism" and "that scientists have added dark matter ad hoc to make equations work. He argued that if the Big Bang is true, the universe must be homogeneous and yet did not explain why that should be true. If anything, Newtonian physics—the law of universal gravitation—says that things should form in clumps as larger masses attracted smaller masses into them."

This was an astute observation. If one posits that at the beginning of the Big Bang (assuming there was a Big Bang) matter or proto-matter was spewed out in all directions, it is hard to conceive how that material could have radiated exactly equally toward all distant points. If there were the slightest disturbance from equal distribution and speed, matter indeed would have begun to "form into clumps." When those clumps became large enough, they would have formed stars and various bodies that orbit stars. Perfect homogeneity is precisely what one would not expect to find. This means that the lack of homogeneity, which can be seen even with the naked eye, is no argument against the Big Bang.

Later in the debate, Cousins brought up the Doppler shift, in which light from stars varies in color depending on whether the stars are moving toward or away from us. Cousins argued that the stable Doppler shifts that are observed by astronomers (mostly the shifts are to red) are proof that, on the whole, other galaxies are moving away from ours. Sungenis, said Flora, "countered the Doppler shift data by citing data from 1932 and citing his book [*Galileo Was Wrong*]. Note that he was not actually providing the evidence from his book, merely promising answers which only could be obtained by purchasing and reading it. Given that he wrote it, I am sadly disappointed that he was incapable of providing the direct evidence verbatim. Another gem: 'Mathematics cannot prove anything.' Ironic, considering that mathematics is the only science that can deal in literal proofs!"

Flora found the question-and-answer period amusing. "An astrophysics postdoctoral student asked Dr. Sungenis to define the dipole, quadrupole, and octopole—something he couldn't do. She also rightly pointed out that of course we are at the center of the observable universe, by sheer definition, since we can see a specific radius around us. Dr. Sungenis countered—utterly failing to appreciate the irony of his statement—that of course she might think that, since she had been indoctrinated over the course of her Ph.D."

THE STORY THUS FAR 5

AT ONE TIME Gerardus Bouw was the most prominent geocentrist in America. Today that title is held by Robert Sungenis. In word count he has written more on geocentrism than have all other living geocentrists combined. He has written a comparable number of words in support of a wide variety of conspiracy theories and against what he thinks are malign influences by Jews in politics, culture, and current events. In his mind, the two categories often become conflated: conspiracies run by Jews.

Sungenis does not subscribe to just one or two conspiracy theories. At his website and in other venues he wrote in favor of many such theories or provided links to the writings of other conspiracists. He has removed those writings and links from his website and has tried to distance himself from his pro-conspiracy and anti-Jewish writings, though not with great success. Instead of disavowing what he wrote, he has said that he no longer wishes to discuss topics that once preoccupied his mind. There has been no change of heart, but there has been a change of tactics. The change was necessitated by his most recent project, the production and promotion of a motion picture.

Sungenis stumbles when trying to deal with the topic of geostationary satellites. He confuses them with GPS satellites, the functioning of which he misunderstands. He offers incompatible explanations of how—assuming a motionless Earth—a geostationary satellite can remain suspended over a fixed spot. This is a particularly awkward issue for him and for other geocentrists, but it is not the only issue on which they cut corners. Often they use scientific terms that they have trouble defining, as shown during a debate Sungenis participated in.

Many geocentrists subscribe to conspiracy theories. (Sungenis does not stand out except in the number of conspiracies he has endorsed.) Whatever else a conspiracy theory may be, it is a shortcut to making sense of a problem. It provides a way to understand something that in reality might have no clean, satisfying answer. For many, a conspiracy theory is a way to get around the hard work of having to think things through.

There is an analogue with the new geocentrists. Instead of working up the physics or mathematics that might establish a particular point, they make grand claims, engage in hand-waving (to use Todd Charles Wood's term), and appeal to Einstein's theories (which they otherwise reject) to say that if something can be demonstrated in heliocentric terms, it can be demonstrated as easily in geocentric terms—even if no one has managed to do so yet.

PART 6

LESS THAN MEETS THE EYE

T HE ESSENCE OF geocentrism is that the Earth is at the center of things—not just at the center of our planetary system but at the center of the entire universe (even if, in the modified Tychonian model, the universe orbits the Sun while the Sun orbits the Earth, which would mean that the universe would not so much orbit the Earth as wobble around it). It is one thing to argue for this centrality based on observations of the Sun, planets, and distant stars. It would be something else to argue also in terms of something between and beyond the stars.

Today's geocentrists believe they have found such a thing in the cosmic microwave background (CMB), a radiation that permeates space. They argue that the CMB, which is thought to be the after-effect of the Big Bang, "points" toward the Earth, which, in their thinking, is located at what had been the Big Bang's point of origin.

It was a Catholic priest, Georges Lemaître, who in 1927 proposed the idea of the Big Bang, though it was not called that until later. His idea found observational support in 1929 when Edwin Hubble noticed a correlation between a galaxy's distance from Earth and the color of the light emitted by the galaxy: the further away, the more heightened was the red shift, which indicates a galaxy that is moving away from Earth. The furthest galaxies were moving away the fastest; the nearest galaxies were moving away the slowest.

Geocentrists' interest in the CMB is not so much in what that radiation may tell us about how the universe got to be the way it is but in where that radiation may point. They think it points toward the Earth, which suggests, to them, that the Earth is the center of all. This is the thesis of *The Principle*, the 2014 motion picture produced by Robert Sungenis and Rick DeLano. *The Principle* is not an overt argument for geocentrism. It approaches the topic through indirection. It does not compare the Ptolemaic, Tychonian, Copernican, and Keplerian systems. It does not feature illustrations of orbits with epicycles. It makes no effort to explain the Galileo case or to argue that Scripture mandates belief in geocentrism. Such things are to be reserved for a planned sequel and for already-existing texts.

(In a December 14, 2013, forum post at Reality Reviewed, a conspiracist website, Sungenis announced that a sequel to *The Principle* is in the works: "Gentlemen, thank you so much for the interest in our movie. . . . We are also making another movie, but it is only for DVD. It will be released concurrently with *The Principle* [this did not happen]. Its title is *Journey to the Center of the Universe* [which also will be the title of a sequel to the book *The Copernican Conspiracy*]. It is a long, detailed, and technical look at the issues brought up in *The Principle*. I am the narrator.

It has a lot of graphics, charts, papers, and such."[106] Reality Reviewed's website has seven main sections. One is about geocentrism. Other sections contain articles that claim that the Moon landing never happened, that far fewer than six million Jews died during the Holocaust, and that Jacqueline Kennedy, under orders from Lyndon Johnson, shot her husband at close range.)

The goal of *The Principle* is modest. Its trailer asks, "Are you significant?" (Everyone wishes to be "significant" in some sense.) The producers suggest the answer truly can be "yes," for individuals and for mankind as a whole, only if Earth is at the physical center of the universe. If it is anywhere other than the center, the answer will be "no" or, at best, "maybe." They think they can demonstrate the affirmative answer by showing that the CMB "points" to Earth.

In *The Principle* Sungenis and DeLano note that certain features of the CMB are aligned with Earth's equator and with the ecliptic, which is the plane of Earth's orbit around the Sun (or, as they see it, of the Sun's orbit around Earth). These alignments are only approximate, though that is not the impression given in the film, which argues that the alignments "point" more or less directly to Earth. They do not. They point neither to Earth nor to anything else. They are indicators of general arrangements within the visible cosmos but that is all. They do not show spatial location, and they are not signposts.

If they truly "pointed" to Earth, a traveler in space could follow them and arrive at least in Earth's vicinity,

106. Realityreviewed.com/phpbb/viewtopic.php?f=17&t=29&sid=c0914d3
4c2afb8ca07fd7c2e4e91bc00.

but that would not happen. It is as though a castaway were in the middle of the ocean. The ocean's apparent flatness is "directional" in the sense that it is a plane and "points" left and right, forward and backward, but not up or down. That flatness helps not a whit in knowing which way to go to make landfall. Much of *The Principle* is taken up with breathy statements about how features of the CMB suggest that man occupies a special place in the universe (perhaps the very center of the universe), but the lengthy argument fails because those features in fact do not "point" to Earth.

MOVIE MEN

SUNGENIS AND DELANO had promised that *The Principle* would be in theaters by the spring of 2014, that it would be handled by a major film distribution company, and that it would open nationwide in multiple theaters. None of that came to pass. The theatrical debut of the film occurred on October 24, 2014, about six months later than predicted. The film opened not nationwide but in a single theater in Addison, Illinois, a thirty-minute drive from downtown Chicago. And *The Principle* was promoted not by a major distributor but by the two proprietors of Rocky Mountain Pictures, a firm that has represented few films, nearly all of them financial failures.

Box office receipts for the opening weekend were $8,657. This put the film's rank at 73 out of 108 films playing nationally at that time. The next time it screened, two weeks later, was at an AMC theater in the Loop. AMC is the second-largest theater chain in the country, and its theater was in a prime location, yet box office receipts for the second weekend were only $7,361. That was a fifteen percent drop even though the film played in a much more prominent location.

Despite the lower revenues, DeLano touted the second weekend as a success: "We are on track for a weekend possibly sufficient to put *The Principle* past the total box office for [*The*] *Unbelievers'* whole run in just two weekends!"[107] This must have struck some as an odd boast. The website for *The Unbelievers* says it "follows renowned scientists Richard Dawkins and Lawrence Krauss across the globe as they speak publicly about the importance of science and reason in the modern world, encouraging others to cast off antiquated religious and politically motivated approaches toward important current issues."[108]

Before going to DVD, where it has had some success, *The Unbelievers* was shown in a single theater for two weeks late in 2013. Total box office receipts were $11,569. So, yes, in a comparable period *The Principle* took in more at the box office, but why compare *The Unbelievers* to *The Principle*? The answer: Lawrence Krauss also appeared in the Sungenis and DeLano film, and, after its premiere, he denounced it and claimed he was duped into being interviewed for it.

Krauss was not the only member of the cast to complain about *The Principle*. Kate Mulgrew's best-remembered role is as Capt. Kathryn Janeway on *Star Trek: Voyager* (1995-2001). Her voice is instantly recognizable to sci-fi fans, which is why she was hired to do the voice-over narration for *The Principle*. When asked to do the work, she was not

107. Rick DeLano (Nov. 8, 2014), facebook.com/rick.delano1?fref=ts.
108. Unbelieversmovie.com/info.htm.

told that the film was intended to lay the groundwork for a pro-geocentric argument. When she discovered the real purpose of the film and who was behind it, she issued this disclaimer at her Facebook page on April 8, 2014:

> I understand there has been some controversy about my participation in a documentary called *The Principle*. Let me assure everyone that I completely agree with the eminent physicist Lawrence Krauss, who was himself misrepresented in the film and who has written a succinct rebuttal in *Slate*. I am not a geocentrist, nor am I in any way a proponent of geocentrism.
>
> More importantly, I do not subscribe to anything Robert Sungenis has written regarding science and history and, had I known of his involvement, would most certainly have avoided this documentary. I was a voice for hire, and a misinformed one, at that.[109]

Sungenis and DeLano defended themselves by noting that Mulgrew had read the script before she did the voice-over narration. This no doubt was true, but a narration is done before a solitary microphone in a small recording studio. The narrator need not be aware of how what is recorded will be set against visuals and against words spoken by other participants in the film. Some of the lines narrated by Mulgrew, as written by DeLano, were hyperbolic (from the trailer: "Everything we think we know about our universe

109. Facebook.com/pages/Kate-Mulgrew-Fan-Page/7122967465.

is wrong"), but nothing in the script Mulgrew was handed could have led anyone to understand how the film would be structured or what its producers hoped it would accomplish regarding their pet theory.

As the date set for the premiere of *The Principle* neared, DeLano began to intimate that the launch might be disappointing. He wrote, "The film will premiere in Chicago on October 24. It will open in one theater. People will come, or they won't. If they do, we will be in more theaters the next weekend. If they don't, we will release the film on DVD and VOD. At this point everything will be determined by our opening weekend in Chicago. If people come, we will be everywhere, in all formats. If people don't come, we will have chewing gum, bailing wire, and a great movie."[110]

DeLano was preparing his and Sungenis's followers for disappointment. It looked like the film that was going to sweep the country might be swept up in a flurry of disinterest. He did what he could to encourage people to bring groups to the opening weekend performances, steering the groups toward the matinee and Sunday showings that otherwise might see small audiences. This seemed to work. For one afternoon showing, for example, the small screening room (which had only 66 seats) sold out days in advance. Naturally enough, DeLano and Sungenis were in attendance at the official debut, which was the 7:00 p.m. Friday showing. That showing was sold out too. So were a few others, but one showing on opening day had only six people in the audience.

In the end, the opening weekend drew fewer than 900

110. Facebook.com/rick.delano1 (Sep. 17, 2014).

viewers. There was every prospect that attendance for a second weekend in Addison would be far smaller—after all, the producers had done what they could to get people to show up for the gala premiere. In the days prior to it, some of their fans said they would be driving from considerable distances to attend. Out-of-towners, if they were coming at all, would come the first weekend. Ditto for the most interested locals. The second weekend promised to draw fewer locals and perhaps no one from a distance. The theater would not make money that way, so the film was not renewed for a second week.

If distributed on a standard contractual basis, a film generally will remain in a theater for at least four weeks because it takes that long for the theater to start making a profit from box office receipts. Until the fourth week, receipts go almost entirely to the film's distributor, with the theater receiving income from concession sales and a weekly "nut" to cover its overhead.

When a film is shown the door after one week, that means either that it drew a much smaller audience than expected and that the theater is trying to cut its losses or—and this is common for independently-produced films—that the engagement was limited to a single week from the get-go, with the distributor paying the theater a flat rental rate for use of the space. That puts all the risk on the distributor and none of the theater, which can look forward to concession sales during the short run. Even if no one shows up, the theater gets the rental fee. If the film does unexpectedly well, the distributor can earn more than the rental cost. Usually films do not come close to covering their costs in such an arrangement, but they may get sufficient attention to make themselves attractive to other theaters.

A few weeks before *The Principle* had its theatrical debut, I had a chance to watch it. The following paragraphs are taken from a review I wrote:

> In terms of production values, the *The Principle* scores well. The graphics are nice, and even the stock footage is well chosen, though the eye tires after a while from too many images of spinning galaxies. The titling is in a good, clean font. The special effects are professional if not remarkable. The music is appropriate to the images and doesn't overpower what is being seen. . . .

> The film is on a par with television documentaries and with programs that have the obligatory visit to Stonehenge (which, yes, appears in *The Principle*). The film's editor took many hours of taped interviews and ancillary footage and made something that holds together for ninety minutes—no small accomplishment.

> *The Principle* is not a drama. It has no characters per se and no plot line. It consists chiefly of talking heads. There are a dozen men who are interviewed in short, quickly alternating snatches. The majority of them are scientists of greater or lesser note who, had they been told what the film was intended to lay the groundwork for (geocentrism) probably

would not have consented to be interviewed. (We know that to be the case with some of them; they have claimed to have been misled about the producers' intentions.)

At the end of the film's credits is a section called "Special Thanks." Among the several dozen names listed are those of Judith Sharpe, promoter through podcasts of Gerry Matatics's sedevacantist theories; Michael King of now-defunct Fisher More College in Fort Worth; E. Michael Jones, editor of *Culture Wars* magazine; Christopher Ferrara, chief writer for *The Remnant*; and Michael Voris, head of ChurchMilitant.tv.

Judith Sharpe's connections to the film and to geocentrism are unclear. She records regular interviews with one-time Catholic apologist Gerry Matatics, who now believes that the Catholic Church has strayed so far from truth that there no longer are any valid bishops and perhaps not even any valid priests left in the world, except for a few who may be in hiding. Sharpe also promotes the work of Hugh Akins, who believes that the *Protocols of the Elders of Zion* are authentic. His book *Synagogue Rising* is about "Israel and Zionist neo-cons in our government and media [who] had their fingerprints all over the 9/11 crime scene." (In this Akins agrees with Robert Sungenis, who has said the same—both about 9/11 and the *Protocols*.)

Michael King has a connection to *The Principle* that is not much less tenuous than Sharpe's. Seven months before his policies drove his small liberal arts college out of business, he had Sungenis speak there. That was in October

2013. As explained at the school's website, Sungenis "lectured on geocentrism, using arguments based on Sacred Scripture, traditional Church teaching, and science that the Earth is fixed at the center of the universe." He "showed lengthy clips of his soon-to-be released movie entitled *The Principle* as part of his presentation."

E. Michael Jones (born 1948) has long written about what he calls "the revolutionary Jew." In a more sophisticated way than Sungenis, Jones sees Jews—or at least a substantial subset of them—as perpetual opponents of Christendom, but he does not seem to endorse any of the conspiracy theories that used to be promoted at Sungenis's website. In his magazine, *Culture Wars*, Jones has not written about geocentrism, so it is unclear why he spoke at the first (and so far only) annual geocentrism conference that was held in South Bend, Indiana, where he resides.

Christopher Ferrara (born 1952) probably has written more words for *The Remnant* than any two or three other writers combined. An attorney, he spends much time complaining in *The Remnant* about the state of the Church and about the motives of people with whom he disagrees. When he is not writing for *The Remnant* he works with Fr. Nicholas Gruner's Fatima Center. It was during a 2013 conference sponsored by Gruner's group that Ferrara taped a half-hour video interview with Sungenis. Sungenis was at the conference to speak about a Fatima-related topic, but the interview was titled "Geocentrism—Crackpot Theory?" At the opening of the exchange Ferrara said that he had been skeptical of geocentrism, laughing it off, but then came across Sungenis's books and found them largely

persuasive. During the interview Ferrara offered no challenge to Sungenis's scientific representations.

An equally supine interview technique was used later
by Michael Voris when he interviewed Sungenis and Rick
DeLano. To promote the launch of their film, the two
appeared twice on Voris's *Mic'd Up* program, which is part of
ChurchMilitant.tv, "the world's first Catholic television sta-
tion exclusively on the Internet." Voris (born 1961) is best
known for a different program, *The Vortex*, "where lies and
falsehoods are trapped and exposed," to use its tag line. He
has developed a reputation as a firebrand who speaks from a
conservative, though not Traditionalist, Catholic viewpoint.
He has complained about bishops and other Church func-
tionaries who fail to ask hard questions, so it was telling that,
when Sungenis and DeLano were interviewed by him, Voris
asked them not a single pointed question about their film,
their scientific theories, or their scientific bona fides.

The first *Mic'd Up* show aired on January 8, 2014, the
second on May 28. In the first, Voris volunteered that he
knew almost nothing about science. In the second Sungenis
and DeLano said that they had in their possession signed
releases from the unbelieving cosmologists interviewed
in their film. This was said in response to allegations by
Lawrence Krauss that he and others had been misled about
the purpose of *The Principle* and that they would not have
consented to be interviewed had the producers been can-
did about their own beliefs. Krauss said he and the other
cosmologists had not been apprised of the film's undergirding thesis—that geocentrism is true—and would not have
consented to participate in a project that, in their minds,
tended to promote scientific error.

The flourishing of the signed releases by Sungenis and DeLano was a dodge, since no one had asserted that Krauss, Michio Kaku, Max Tegmark, and the other interviewees had not signed releases in exchange for modest honoraria. Voris failed to pick up on this. He did not ask Sungenis and DeLano how open they had been about the purpose of their film.

During the *Mic'd Up* shows Sungenis said little. The majority of the words were spoken by DeLano, with Voris a distant second. The two of them posited the existence of a concerted effort to undercut the film by what Voris elsewhere had dubbed "the Church of Nice." Voris, with a chuckle, brought up the charge of anti-Semitism to Sungenis, who denied harboring animosity toward Jews. Neither made any reference to the tens of thousands of words Sungenis had written against Jews since 2002. Sungenis claimed that he had criticized Catholics twice as often as he criticized Jews. Voris threw him two slow pitches: "Are you a Holocaust denier?" "No." "Do you hate Jews?" "No." That was as probing as Voris got.

(Sungenis does not say that the Holocaust never occurred but that the number of Jews who met their deaths during World War II was very much smaller than is commonly thought and that most of the victims died from disease fostered by insanitary conditions in the camps. He argues that "records of the Red Cross" demonstrate that "the numbers were in the hundred thousands, not millions, and it documents that most Jews died of disease." To an inquirer who thought he was saying the Nazis did not put Jews in camps, Sungenis replied: "The issue here is whether the Germans gassed the Jews, not interned them. No one disagrees that the Jews were interned. But there is no evidence that the Jews were gassed.")

In a separate program, Michael Voris's associate Christine Niles added to the promotion of *The Principle*. She conducted with DeLano what David Palm termed "an infomercial/interview." Palm said, "Depending upon whom he's talking to at the moment, DeLano can be coy about the ultimate intent behind the film. But in this interview Niles and DeLano make it very plain that geocentrism is first and foremost a matter of faith, not a matter of science."

As was the case with Voris in the two *Mic'd Up* episodes, Niles asked not a single probing question and challenged none of DeLano's scientific ideas. Niles said she is not a geocentrist. One might think that she would hold much of what DeLano believes to be just plain wrong—bad science and bad theology—yet she handled him as though she subscribed to geocentrist positions.

THE MAN BEHIND THE CURTAIN

I N THE END-OF-FILM credits for *The Principle*, Robert Sungenis is listed as executive producer, a title indicating not that he oversaw the business side of production but that he was responsible for procuring funding. The producer is listed as Rick DeLano, who also is listed as writer of the screenplay.

Not much is known about DeLano, who was born in 1957. On the tax return of Sungenis's Catholic Apologetics International DeLano has been listed as one of four directors. His residence address is a boat slip at the Port Royal Marina in Redondo Beach, California. He claims to have had wide experience in the entertainment industry but has provided no specifics. He once described himself as "the father of two sons of a Jewish mother"[111]—this said in the context of a defense of Sungenis against charges of anti-Semitism. DeLano's name originally seems to have been spelled Delaneaux; there is no indication when or why the spelling was changed.

Regarding his education, DeLano said, "I never

111. E-mail to the author (Oct. 19, 2006).

attended high school. I did attend college, but only until I achieved the age of 15 years and 9 months, at which point the statutory requirement of compulsory education ceased in my home state. I returned to college somewhat later. . . . I happily departed academia, never to return."[112]

When Catholic apologist Dave Armstrong asked him which science courses he had taken, at first there was no reply. Armstrong wrote that DeLano "evaded my question about his science education four times, in two different highly visible venues."[113] Later DeLano added some information: "I spent five years researching the relevant science [about geocentrism], and it pays off in the interviews [in *The Principle*]. I took the general science classes at my primary and middle schools."[114] Those classes, completed when he was no older than about fourteen, apparently were the extent of his schooling in science.

DeLano has seemed to welcome any publicity that has come his way, the better to promote *The Principle*, but there is one write-up that he has avoided bringing to people's attention. That is an article that appeared on the website of *Popular Science* magazine on May 7, 2014. The title was "The Conspiracy Theorist Who Duped the World's Biggest Physicists."[115]

The writer, Colin Lecher, explained to his readers that

112. This remark was made in an online exchange with Catholic apologist Dave Armstrong (Jan. 8, 2014), socrates58.blogspot.com/2014/01/formal-science-education-of-rick-delano.html
113. Ibid.
114. Ibid.
115. Colin Lecher, "The Conspiracy Theorist Who Duped the World's Biggest Physicists," *Popular Science* (May 7, 2014), popsci.com/article/science/how-conspiracy-theorist-duped-worlds-biggest-physicists.

THE MAN BEHIND THE CURTAIN

The Principle "promotes the long, long-debunked idea that the Earth is the center of the universe. When it started getting attention in April, *Raw Story*, *Slate*, the *Washington Times*, the *Huffington Post*, and countless others wrote about the film. It was a hit, for all the wrong reasons." This intrigued Lecher, so he contacted DeLano. During their first telephone interview they mainly discussed astronomical theories. At one point Lecher asked DeLano how the film was funded—"a question he immediately dismissed as irrelevant."

After that conversation Lecher learned that DeLano is "the proprietor of a blog, Magisterial Fundies, which focuses on his theories and [on] *The Principle* specifically. . . . He posts about recent—and legitimate—scientific discoveries, then uses them to bolster his ideas, which, after reading more of the blog, take on a Catholic hue. . . . The About section on the blog shows a small, pixelated photo of a man—DeLano, presumably. You can only barely make out bushy eyebrows, a suit jacket, and a smile."

Then Lecher stumbled across something intriguing:

Court records for DeLano turned up one unexpected hit: in 2002 a Rick DeLano was listed as a defendant in a $10 million suit alleging he and others had participated in a scheme to misrepresent stock in Internet companies. "Defendants Perlman, DeLano, and Levy introduced Plaintiff to several individuals whom they claimed were officers and directors of these fifty-four companies ('Companies'). Plaintiff alleges that these representations were intentionally and willfully misleading," according to the records.

In the suit DeLano is listed as a California resident; his current phone number has an area code that puts him in California, too, which would go some way toward explaining his relationship with the film production company. The case settled for an undisclosed amount.

"The week after Easter," said Lecher, "I called him again, and though he didn't answer the first time, he eagerly returned my call. Compared to the last time we spoke, he was lucid and forthcoming, willing to expand on his theories and his film. He told me he was determining 'who the fair-minded media contacts are' and would divulge new information to them accordingly, perhaps even let them interview the director. He clarified that the goal of the film, in his view, was to consider geocentrism as one competing theory among many."

Then everything changed. "But when I asked what he could tell me about the company listed in the lawsuit I had turned up, the chat suddenly bottomed out. 'That would have nothing to do with my film, and I think this conversation is over,' he said. 'Thank you very much.' Then—*click*—we were done."

Lecher noted that DeLano said he might arrange for an interview of the director of *The Principle*. That was Katheryne Thomas, whose only other directorial stint was for the 2007 video *Scum of the Earth—Sleaze Freak: Behind the Scenes*. That film used the tag line "Full of shock-by-shock sleazy confessions." The twenty-minute documentary was about a heavy metal band called Scum of the Earth and the production of its album *Sleaze Freak*, which sold only 800 copies its first week and failed to make the Billboard 200. Perhaps the song titles had something to do with it: "Devilscum," "Death Stomp," "I Am Monster," "Corpse Grinders," "Scum-O-Rama," and so on.

Rick DeLano has not explained why Thomas was chosen to be the director of *The Principle*. She was not the first choice. In a February 2011 letter sent to donors to his apostolate, Catholic Apologetics International, Robert Sungenis said, "We have already drawn up a contract with an experienced and talented Hollywood director. Her name is Nicole Torre. You can look her up on the Internet." Torre's biographical squib, given at Internet Movie Data Base, reports that her "debut documentary feature *Houston We Have a Problem* has played around the world and received a 2011 Emmy nomination." She has several minor awards to her credit and has worked with several television series. She is described as working in "progressive media" and as being "actively involved in several environmental causes. She produced a movie for Physicians for Social Responsibility, an organization involved for five decades in various left-wing causes. In terms of likely political and cultural attitudes, Torre seems much different from Sungenis. Then again, so does Katheryne Thomas.

That fund-raising letter mentioned Torre only in passing. Its chief purpose was to inform Sungenis's donors about his plans for what would become *The Principle*. He noted that he had "received seed money of $150,000 from two dedicated patrons so that we can produce a film documentary on geocentrism. My patrons are convinced, as I am myself, that modern science has conclusively shown that the Earth is motionless in the center of the universe and that the public at large has had this information kept from them for many years."

Sungenis said he and his associates (by which he apparently meant Rick DeLano) had chosen as their film's title "*Journey to the Center of the Universe,*[116] a take off from Jules Verne's *Journey to the Center of the Earth.* The script and story line are registered with the Hollywood Writer's Guild. Our attorney [Jeffrey C. Foy] has established a limited liability company for us called Stellar Motion Pictures, which is located in [West] Hollywood [at Foy's apartment building]. We have the foundation and are ready to go into production."

Then Sungenis began the financial buildup. "Hollywood locals are very interested in our documentary, not necessarily because they have suddenly adopted a Christian view of the universe (although our director [Nicole Torre] has become convinced) but because they believe this will be a big money maker. They know in their gut this is a big story, one of the biggest they've covered in a long time; probably the biggest science story in the last 500 years since Galileo's trial before the Catholic Church."

"These Hollywood people" had come to see, as Sungenis had been insisting, that mainstream science had been hiding the truth about cosmology. He said that "once this cover-up is exposed, the whole world will want to know about it (pro or con) and they will make a lot of money by

116. As mentioned in the chapter "More Wrong than Galileo," at another time the film's tentative title was *Not by Science Alone: Modern Science at the Crossroads of Divine Revelation.*

packing them into theaters to see it in full and fabulous drama." It will be the "Hollywood people" who will become "the very vehicle God chooses to show the world that the Catholic Church was right all along and that the world has been living in an illusion."

How important would Sungenis's documentary be? He saw only two possibilities. First, "the mindset of the whole world will change as we generate new reverence for the Church and for divine revelation, which will produce a spiritual revival accompanied by many conversions to the faith." The other possibility is that "Jesus is about ready to come back and judge the world, and our movie about the Earth in the center of the universe will be a testimony to them in the same way that Noah's boat was a testimony of the end of the world in his day." No modest expectations here. "Whatever the case," said Sungenis, "God plans on using our movie to bring his message to the world in a dynamic way."

Of course, there was a contingency: money. The film could not be produced without substantial cash, but the rewards for participation could be great. Sungenis cited examples of successful documentaries. *Expelled: No Intelligence Allowed* (2008), "cost $3.8 million to produce but it made about $15 million in domestic and overseas sales." *Fahrenheit 911* (2004) "cost $6 million to produce but grossed $250 million." *An Inconvenient Truth* (2006) costs a mere $1.5 million but brought in $75 million.

Sungenis expected his film to "make something between these figures. Once we make that money, we intend on using it to make other movies in this same genre (e.g., the true history of the Galileo affair; the fallacies of evolution and the evidence for creationism; a dramatic movie about a scientist who discovers geocentrism)." Reaching such goals depended on raising funds for the first film. There were procedures to be followed.

"Normally an investor is required to have a million dollars in net worth [and] we must take investments in units of $33,000." The return on investment would be a function of how much capital was raised. "Let's say, for example, you invested $33,000 and the total pool of investment money contributed is $500,000. This amount will allow for a 40 percent return for all the investors"—that is, investors would be entitled to 40 percent of the profits. "And let's say the movie makes a $10 million profit. We then take 40 percent of $10 million as the total amount of profit for the investors, which is $4 million. Since your $33,000 is 6.6 percent of the $500,000 in the initial investment pool, you would receive 6.6 percent of $4 million, or, $264,000. . . . Of course, if our movie made $20 million, you would make $528,000," which equals sixteen times the original investment.

This offer would not remain open indefinitely. "Sometime before mid-year 2011, however, we will have to close the investment opportunity. The rules allow only a certain window to gather potential investors, and once that window closes it cannot be opened again." Sungenis said he was "offering [this opportunity] to you because you have been so generous to me over the years with my apostolate, CAI, and I want to return the favor to you. Also, if you have any friends or relatives who you want to see make a windfall, please let them know of this opportunity."

It is not known whether Sungenis's offer to "make a windfall" resulted in donations for the film project, but three months later, Stellar Motion Pictures filed Form D with the Securities and Exchange Commission. The form must be filed each time a privately-held company wants to raise capital through an exempt offering of securities. The first Form D was filed on April 7, 2011. The company's offering amount was $2.5 million. Nothing was sold. A year later, on April 5, 2012, that Form D was amended, and $343,000 was raised from eleven investors, who are not named in the form. The third Form D was filed on June 27, 2013, with an offering of $1.7 million, but none of that was sold. Of the $343,000, $10,000 went to sales commissions and $85,000 to "officers and directors of the company."

In an online exchange Sungenis said that he raised nearly one million dollars to underwrite the project,[117] but it is not clear how that should be taken. Was that amount raised from investors, from donors to his apostolate, or in part from his own funds? Was that money cash in hard or in the form of promissory notes?

The Principle is produced by Stellar Motion Pictures, LLC, which, according to Form D, has three "key executives," each of whom is denominated "executive director." They are Sungenis, DeLano, and Jeffery C. Foy, who describes himself at his personal website as "an entertainment lawyer, writer, and independent producer." The West Hollywood address for his law office, which also is the legal address for Stellar Motion Pictures, is an apartment building. At his website Foy has described *The Principle* as a "fringe science feature documentary."[118]

117. Realityreviewed.com/phpbb/viewtopic.php?f=17&t=29&sid=a7bed53
63ca3e6a01377683e6e924c7e. Reality Reviewed is a conspiracist website.
118. Blogger.com/profile/03918487647629187584.

UNPRINCIPLED

J OHN HARTNETT ONCE was a Catholic but left the Church for Protestant Fundamentalism: "what freedom did that bring." Responding to remarks made by Pope Francis to the Pontifical Academy of Sciences in October 2014, Hartnett said, "I hold God's Word to be true over what any man or Jesuit might teach. He [Francis] has now declared (again) Big Bang and evolution are true history of the world. No way. I will stick with the paper pope [the Bible]."

An Australian born in 1952, Hartnett might seem a peculiar choice as ally to the Catholic geocentrists who produced *The Principle*, particularly to Robert Sungenis, who, in the 1990s, gained notice as a Catholic apologist by writing against the sort of anti-Catholicism sometimes shown by former Catholics such as Hartnett, but, like politics, cosmology makes strange bedfellows. Hartnett is valuable to the geocentrist cause because he is one of the few proponents who can boast a Ph.D. in physics. He has bona fides that other geocentrists do not have, but his status comes with a price. He is a geocentrist in a distinctive way.

Hartnett posted a review of *The Principle* at his website on November 3, 2014, about two weeks after the film's theatrical debut. He said, "The planet Earth is in a special place in the universe, which is not the same thing as being absolutely geocentric."[119] His position is that Earth is somewhere near the center of the universe but not actually at its center. In this he differs from Sungenis and Rick DeLano, who hold that Earth literally is at the center of the universe.

Hartnett said he agreed to be interviewed for *The Principle* because he "wanted to present a *biblical creationist non-geocentric point of view*, which still permits our planet Earth to be in a special place in the universe" [his emphasis]. In the film he expresses his "galacto-centric or near galacto-centric worldview": Earth is "somewhere near the local center of a spherically symmetric universe." Earth is "cosmologically near the center of a massive superstructure of galaxies" and is within 100 million light-years of the center of the universe, which Hartnett does not otherwise pinpoint. For him, this nearness to the center is enough to establish that Earth is "in a special place"—but so, then, are billions of other celestial bodies.

Hartnett is not satisfied with the way his ideas were portrayed in *The Principle*. His words were preserved, but they were tied to animation sequences that were contrary to his argument. "The impression [the film] gives the viewer is that [his view] implies the Earth is at some absolute center. I did not say that. The impression is wrong." When, in the film, Hartnett discusses a large-scale structure that suggests

119. John Hartnett, "Review of 'The Principle'" (Nov. 3, 2014), johnhartnett.org/page/2.

that the Earth is near the center of a "massive superstructure of galaxies," the animation shows a solitary Earth at the center of the image, not a galaxy or cluster of galaxies. This happens twice.

The Principle refers to Hartnett's conclusion that the universe shows "concentric shells of galaxies, derived from [his] analysis of large-scale structures in two large galaxy surveys," but, says Hartnett, the film suggests that "the concentric shells are centered on the Earth. Sungenis is cut in to give the impression that I mean an absolute geocentric universe. He says, 'So every phenomenon that you see out there is all on concentric shells around where? The Earth.' Then at the 53:21 mark a picture of the planet Earth is shown, reinforcing the false idea. . . . The 'onion' illustration was my idea. I am sure I explained I was not suggesting the shells were centered precisely on the Earth, but that did not survive the editor's cut."

Hartnett makes an additional complaint about Sungenis, who in the film says "the whole cosmic microwave background is pointing to us, this tiny little point in space, and this is where we live." Hartnett objects. He says the CMB "does not define a unique center. It does not, nor does it point to Earth. It defines an axis in the cosmos, an anisotropy axis. That is a special direction . . . But it does not point at the planet Earth. After that [in the film] the impression is given of a conspiracy at NASA. Oh well, then it must be true." Here Hartnett refers to Sungenis's comment, which is shown also in the film's trailer, that "you can go on some websites of NASA and see that they've started to take down stuff that might hint to a geocentric universe."

In a reply to Hartnett's review of *The Principle*, Sungenis wrote to Hartnett: "It puzzles me why you are so frightened of geocentrism." He said, "Scripture teaches geocentrism, without a doubt. It does so in many places, and in many different ways, from the historical narratives of Genesis and Joshua to the historical poetry of the Psalms and beyond. In fact, God uses the Earth's immobility as a direct analogy to his own immutability."[120]

Sungenis concluded by saying, "I think the reason we can't hold hands is because your people have a higher commitment to General Relativity than they do to the literal words of Scripture, in addition to the fact that having to defend geocentrism would cause undue embarrassment when one is trying to create a credible scientific profile to fight evolution."

Hartnett, the Catholic-turned-Fundamentalist, is not as literal in his beliefs as is Sungenis the Catholic. Writing to a third party, Hartnett said of Sungenis that "I know enough to know that he misapplies Scripture and interprets it contrary to the original intention. . . . If you are going to apply Scripture which says, inter alia, 'the Earth does not move' as meaning it as an absolute anchor point in the universe, immovable, then you must also do that when Scripture speaks of the believer not being moved (Psalm 112:6: 'Surely he shall not be moved for ever'). Scripture must be interpreted consistently in this case also."[121]

Hartnett's reservations about Sungenis go beyond Scripture. They extend also to Sungenis's claim that

120. Facebook.com/john.hartnett.908/
posts/1501280056793351?comment_id=1501587813429242&offset=0&t
otal_comments=31.
121. E-mail from John Hartnett (Jun. 5, 2014).

anything that can be expressed mathematically in terms of a heliocentric coordinate system can be expressed with equal ease in terms of a geocentric coordinate system.

Hartnett demurs. "Of course one can write equations of motion in geocentric coordinates. It is a matter of choice, but they are horribly complicated. . . . In physics it has long been known that the choice of coordinates that simplifies the equations of motion the most gives the clearest description of the physics, which helps us understand what is going on. That is the best choice, and God led Copernicus, Galileo, Kepler, and Newton to give us the best and clearest understanding of the physics." Such a comment might be said to make Harnett a geocentric heliocentrist: someone who believes that the Earth is "in a special place" without being at the center of the universe.

UPSIDE DOWN HISTORY

T HE TAG LINE for *The Principle* is "Are you significant?" Is man simply a cosmic accident, as evidenced
by his existence on a nondescript planet located in a
nondescript solar system far out on one of the spiral arms
of the Milky Way? Or does he have an importance implied
by his living on a planet located at the very center of the
universe?

Geocentrists argue that heliocentrism demoted man
from his God-given status. What had been accepted for
centuries—that man is the pinnacle of God's material creation—was undermined when science taught that man,
like the planet he lives on, is of no particular significance.
This was a revolution in man's outlook on his world. As the
Britannica Concise Encyclopedia puts it, "Dethronement
of Earth from the centre of the universe caused profound
shock: the Copernican system challenged the entire system
of ancient authority and required a complete change in the
philosophical conception of the universe."[122]

Or did it? Is that really what happened, or is that

122. "Copernican System," *Britannica Concise Encyclopedia* (2007).

another Whig interpretation of history, a retrojection into the past of recent prejudices? This issue is dealt with by Dennis R. Danielson, professor of English at the University of British Columbia, in his contribution to *Galileo Went to Jail*, a book that refutes myths of history. Danielson's piece is titled "Myth 6: That Copernicanism Demoted Humans from the Center of the Cosmos."[123]

He begins with "the cliché that geocentrism equates with anthropomorphism." This actually is a modern attitude. Today's writers imagine that the ancients thought that man's position at the center of the universe reflected his elevated status, but Danielson argues that it was quite the reverse. We'll get to this in a moment. Danielson brings up a separate point: "The other assumption that generally piggybacks on the cliché is that Copernicus, in allegedly reducing the status of the Earth, also struck a blow against religion, particularly the Abrahamic religions, which supposedly require the cosmic centrality of humankind. The weaknesses of this view . . . include a failure to distinguish figurative from literal centrality."

If this notion of man's physical centrality were true, one would expect that the issue of man's "dethronement" would have played a key role in Galileo's trial, but it did not. The Galileo affair concerned biblical interpretation "and the threat Galileo represented to the entrenched 'scientific' authority of Aristotle." It had little to do with "any supposed Copernican depreciation of the cosmic specialness or privilege of humankind. If anything, Galileo and his

123. Dennis R. Danielson, "Myth 6: That Copernicanism Demoted Humans from the Center of the Cosmos," in Ronald L. Numbers, *Galileo Goes to Jail* (Cambridge: Harvard University Press, 2009), 50–58.

fellow Copernicans were raising the status of Earth and its inhabitants within the universe."

This is something the new geocentrists, to a man, seem not to appreciate. They seem not to have come across the idea at all. They show little appreciation of how people of the late sixteenth and early seventeenth centuries understood natural philosophy. The system then in place derived from Aristotle (384-322 B.C.), who said the Earth was at the center of the universe because the center is where heavy things aggregate. Of the four natural elements—earth, water, air, and fire—earth (solid matter) is the heaviest and so tends to sink to the center or lowest point of the universe. The lighter elements tend to settle in higher regions, with fire being the furthest from us—witness the Sun and other celestial bodies.

Following Aristotle's lead, Thomas Aquinas (1225-1274) said that "Earth—that all the spheres encircle and that, as for place, lies at the center—is the most material and coarsest [*ignobilissima*] of all bodies."[124] Dante continued this view. In his *Commedia* he placed hell at the center of the Earth, the most ignoble place of all, and heaven beyond the outermost sphere of the stars. Earth's ignobility was confirmed by Giovanni Pico, who wrote that Earth occupied "the excrementary and filthy parts of the lower world,"[125] and by Michel de Montaigne, who said man is "lodged here in the dirt and filth of the world, nailed and

124. Thomas Aquinas, *Commentary on Aristotle's De Caelo* (1272s.), II, xiii, 1.
125. Giovanni Pico della Mirandola, "Oration on the Dignity of Man," in *Renaissance of Man*, Ernst Cassirer, ed. (Chicago: University of Chicago Press, 1948), 224.

riveted to the worst and deadest part of the universe, the lowest story of the house, the most remote from the heavenly arch."[126]

Danielson summarizes by saying that "pre-Copernican cosmology thus implied not the figurative, metaphysical centrality—the importance or specialness—of Earth but rather its physical centrality and at the same time sheer grossness." The center thus was not the best place to be but the worst. Man, living on the surface of the world, was very near hell and very distant from heaven—a testament to his own low status.

What Copernicus ended up doing—this was a consequence of his theories, not a motivating purpose behind them—was to elevate man by removing him from the lowest place, the sump of the universe. The Sun now occupied that spot, yet it did not suffer from its repositioning. Change came not to the Sun but to the new place it occupied. The center "was transformed into a place of honor."

This, says Danielson, "was a remarkable feat: Copernicus thus simultaneously enhanced the cosmic status of both Earth and Sun. And the latter part of his task succeeded so well that, ever since, the Earth's removal from what became the Sun's place of honor has appeared to some as a diminution of its value." But this is a retrojection from our own time into the consciousness of the seventeenth century. "A refutation of that anachronistic interpretation, however, can be put quite simply: for Earth to be raised up out of what were then considered 'the excrementary and filthy parts of the world'

126. Michel de Montaigne, "An Apology for Raymond Sebond," in *The Essays of Michel de Montaigne*, Charles Cotton, trans. (London: George Bell, 1892), 2:134.

can't seriously be interpreted as a demotion. . . . Only with the abolition of geocentrism, then, might we truly say that we occupy the best, most privileged place in the universe"— or, at least, a much better place.

Danielson says it is "hard to know exactly what gave rise to the myth of religious opposition to the supposed 'dethroning' of the Earth." What he calls the "Copernican cliché" apparently "appeared for the first time in France more than a century after the death of Copernicus as part of an anti-anthropocentric critique." He cites Cyrano de Bergerac as perhaps the best known promoter of this opinion but Bernard le Bouvier de Fontenelle's *Discourse on the Plurality of Worlds* (1686) as perhaps the most influential.

In any case, this became the accepted viewpoint of the Enlightenment, "as magisterially summarized in 1810 by Johann Wolfgang [von] Goethe, who repeated the notion that 'no discovery or opinion ever produced a greater effect on the human spirit than did the teaching of Copernicus,' for it obliged Earth 'to relinquish the colossal privilege of being the center of the universe.'"[127]

It is this Enlightenment misinterpretation that today's geocentrists have adopted, without knowing the provenance of their opinion. They think the ancient system, which had Earth at the center of the universe, was understood by its adherents as giving Earth and, therefore, man a status of honor, and they think that the heliocentric position destroyed that status. They have it backwards. Lowly man had been in the lowest and most ignoble place in the

127. Johann Wolfgang von Goethe, "Materialien zur Geschichte der Farbenlehre," in *Goethes Werke* (Hamburg: Christian Wegner Verlag, 1960), 14:81.

universe (aside from hell itself, which was lower and more ignoble than was the surface of the Earth). He had been as far away as possible from the empyrean reaches of heaven, beyond which God dwelt.

It had not been Copernicus's intention to lower or elevate man. His interests were not so much with the metaphysical as with the physical. The same can be said for Galileo and Kepler. They were neither trying to prove nor disprove a theological point. What they ended up doing, almost by accident, was moving man a step up in the chain of being. He was not, by any means, much closer to heaven, but he was a few steps removed from being next door to hell.

THE STORY THUS FAR 6

FOR GEOCENTRISM TO triumph, heliocentrism must be undermined. A frontal assault would accomplish little, since heliocentrism has been entrenched for four centuries, and only the tiniest fraction of one percent of scientists reject heliocentrism. But mainstream scientists do have something to offer geocentrists: uncertainty, particularly with respect to cosmic origins. Today's cosmologists admit that their field is in flux. Theories are changing as telescopes and satellites gather ever more precise data. Physicists and astronomers find themselves unable to provide answers to some questions that are put to them. In some areas their ideas are more unsettled today than at any time since the Special Theory of Relativity was formulated in 1905.

This gives geocentrists an opening. They are able to suggest that scientists' present inability to explain certain phenomena and their reliance on theoretical constructs that as yet have not been proved (such as the existence of dark matter) mean that mainstream science is at a dead end. Dead ends are reached when one has taken the wrong route. The solution is to go back and rejoin the correct route that goes by way of geocentrism.

Leading geocentrists have chosen cosmic microwave background as the phenomenon on which to hang their claim of a wrong turn. They have found prominent scientists who acknowledge surprise in discovering that the CMB displays an alignment. While that alignment may be said to "point" in a general direction, it doesn't point at anything in particular, but geocentrists disingenuously insist it points at Earth. They conclude Earth is at the center of the universe, perhaps occupying the spot from which the Big Bang emanated (assuming the Big Bang occurred at all).

While no mainstream scientist agrees with this thinking, it is the thesis around which Robert Sungenis and Rick DeLano have built a film. They managed to get interviews with top scientists, none of whom was aware that he was being interviewed for a documentary intended to lay the groundwork for geocentrism. Even one of the sympathetic interviewees, John Hartnett, has distanced himself from *The Principle*, saying it misrepresents his ideas.

The overarching supposition of the film and of modern geocentrism is that heliocentrism has been the remote source of many of today's woes. By displacing the Earth from the center of the cosmos, heliocentrism reduced not only Earth's importance but man's. If the universe has no center—or even if it does but Earth is not located at it—then Earth becomes an inconsequential speck in the empty vastness of space, and man becomes little more than a biological accident. Yet man is no accident. He is created in the image and likeness of God and therefore truly is special. To the geocentrist this implies that man's location also must be special.

This argument can be made only by someone who misunderstands the ancient status of a central Earth. Far from being the most honored place, the center was thought to be the most ignoble. After all, hell itself was said to be not far beneath men's feet. When Copernicus put the Sun at the center of our planetary system, he elevated the status not only of Earth but of the center itself. This was not his intention, but such was the result of his astronomical theory.

This means *The Principle* is based on a misapprehension. It goes wrong from its very premise, and it goes wrong when it tries to read into uncertainties in cosmological theory a necessity to return to geocentrism. To whatever extent the film succeeds, its success will be a function not of the scientific acumen of its producers but of the capacity of its audience to be swayed by assertions that seem true but are in fact false.

CONCLUSION

O N DARK MEDITERRANEAN nights the ancients observed the movements of the planets and stars, and during the day they observed the movement of the Sun. They noticed general regularity marked by particular irregularity. Movements shifted slowly during the year, the Sun rising and setting farther to the north, then farther to the south, the planets swimming among the stars, racing one another through the constellations, sometimes overtaking one another, sometimes apparently moving backward. Even the stars themselves processed through the skies.

Since man is incapable of not drawing inferences from observations, the ancients guessed at the arrangement of heavenly bodies. Some surmised that the Sun, like a candle quenched in a bowl, was extinguished each evening as it was swallowed by the sea. Others thought this improbable and speculated that the Sun circled behind the Earth at night. Even to the unsophisticated, the planets and stars seemed ordered; perhaps they too, like the Sun, circled the Earth. Thus there arose astronomical hypotheses.

Proto-astronomers, such as the Wise Men of Matthew's Gospel, went further, carefully measuring the celestial movements, keeping records of the rising and setting of the Sun, the journeys of the planets, the positions of the stars. The accuracy of their measurements was limited by the power of the naked eye and the absence of chronometers, yet they adduced evidence, applied it to the prevailing hypotheses, and ended up with theories, which are hypotheses united with evidence that tends to support them.

The longest-lasting and most influential cosmological theory was that of Ptolemy, who, relying on earlier observers such as Hipparchus and Timocharis, developed a geocentric model that seemed to account for all observed movements—as the medievals later put it, it "saved the appearances." For centuries few doubted that the Earth stood at the center of what today we call the solar system.

But the Ptolemaic theory, raised from mere hypothesis through the application of scientific measurements, proved in the end to be false. The Sun, not the Earth, is the center of the solar system; the planets move along ellipses, not along cycles and epicycles; the stars, so distant that their movements are almost imperceptible, do not circle the Earth but are, for practical purposes, fixed.

Thus it is in science. Initial observations produce hypotheses, which are mere guesses, some immediately seen to be improbable (such as that the setting Sun is extinguished each day), some seen to be possible (such as that the Sun, planets, and stars circle the Earth). Through scientific investigation and the gathering of innumerable tiny facts, scientists come to single out one hypothesis and produce an overarching explanation that accommodates the evidence. This explanation is called a theory.

A theory is not the same as a truth. A scientific theory

may be true, or it may be false. It may be both true and false. A theory is drawn from facts that, individually, are thought to be true, but it is more than a mere agglomeration of facts. It is an overarching explanation of why those facts are as they are and what those facts mean, but even the most convincing or satisfying theory is tentative. Any theory stands in danger of being supplanted by a better one.

Sometimes a new theory is supplanted by a modified version of its predecessor. A theory thought to be dead is revived in slightly different dress. It might be because the new theory, once thought so compelling, proves to be unworkable, necessitating an about-face, or it might be for reasons having little to do with the intrinsic worth of the old theory. The latter has been the case with geocentrism. It has made a small comeback, but its attraction is not really in its science.

The new geocentrists may say their abandonment of Copernicus, Kepler, Newton, and Einstein stems from weaknesses within the heliocentric theory, but in fact the return to geocentrism is based upon a particular style of scriptural interpretation. Were it not for an unbending literalism, the new geocentrism might have no proponents. The scientific case for it, such as it is, has been worked up to accommodate a particular interpretation of the Bible.

Will this resurrected cosmological theory gain a larger following, becoming a long-term thorn in the side of the scientific establishment, or will it disappear once time takes its current proponents from the scene? The answer will not be known for several decades, but what can be known today is that the new geocentrists believe themselves to be heralds of old truths made new, called to spread their gospel far and wide.

ACKNOWLEDGEMENTS

I wish to thank Michael Forrest, Alec MacAndrew, and David Palm for reviewing a draft of *The New Geocentrists*. Their corrections and suggestions made the text better than it otherwise would have been. Any remaining errors and imprecisions are mine, not theirs.

www.ingramcontent.com/pod-product-compliance
Lightning Source LLC
Chambersburg PA
CBHW032033080426
42733CB00006B/68